The Caretakers of the Cosmos

Also by Gary Lachman

Madame Blavatsky: The Mother of Modern Spirituality

Swedenborg: An Introduction to His Life and Ideas

The Quest for Hermes Trismegistus: From Ancient Egypt to the Modern World

Jung the Mystic: The Esoteric Dimensions of Carl Jung's Life and Teachings

The Dedalus Book of Literary Suicides: Dead Letters

Politics and the Occult: The Left, the Right, and the Radically Unseen

Rudolf Steiner: An Introduction to His Life and Work

The Dedalus Occult Reader: The Garden of Hermetic Dreams

In Search of P.D. Ouspensky: The Genius in the Shadow of Gurdjieff

The Dedalus Book of the Occult: A Dark Muse

A Secret History of Consciousness

Turn Off Your Mind: The Mystic Sixties and the Dark Side of the Age of Aquarius

Two Essays on Colin Wilson

As Gary Valentine:

New York Rocker: My Life in the Blank Generation, with Blondie, Iggy Pop, and Others 1974-1981.

The Caretakers of the Cosmos

LIVING RESPONSIBLY IN AN UNFINISHED WORLD

GARY LACHMAN

First published by Floris Books in 2013
© 2013 Gary Lachman

Gary Lachman has asserted his right under the
Copyright, Design and Patents Act 1988
to be identified as the Author of this work

All rights reserved. No part of this book may be
reproduced without the prior permission of
Floris Books, 15 Harrison Gardens, Edinburgh
www.florisbooks.co.uk

 This book is also available as an eBook

British Library CIP Data available
ISBN 978-178250-002-5
Printed in Great Britain
by Bell & Bain Ltd, Glasgow

Contents

Acknowledgments	9
Introduction: Saving the Universe	11
1. The Other Side	29
2. A Dweller on Two Worlds	56
3. Doing the Good that You Know	90
4. The Good Society	116
5. Beyond Nature	142
6. The Participatory Universe	171
7. As Far as Thought Can Reach	206
Notes	225
Selected Bibliography	243
Index	249

*For Colin Wilson, who has certainly repaired
quite a bit of the universe.*

If we all desire it intensely, if we organise all the visible and invisible powers of earth and fling them upward, if we all battle together like fellow combatants eternally vigilant – then the Universe might possibly be saved.

Nikos Kazantzakis

Acknowledgements

Many people have helped make this book possible. I'd like to thank my editor Christopher Moore for suggesting it and Floris Books for being supportive. Thanks too go to the staff of the British Library for their ever reliable help. My friend James Hamilton endured many conversations and made many suggestions. Special thanks goes to Iain McGilchrist for reading some of the work while in progress and for his helpful comments, and Anja Flode Bjorlo for crucial help with the index and research. My friends John Browner, Lisa Yarger, and their daughter Greta were once again extremely generous in their hospitality in Munich, where some of this book was written; aptly, a chapter having much to do with Max Scheler, who was born there. As always my sons Maximilian and Joshua and their mother Ruth Jones were of inestimable assistance in providing me with some very special reasons for taking care of the cosmos.

Introduction: Saving the Universe

This book has a bold title, and it may be a good idea to begin by trying to explain it. While working on an earlier book, *The Quest for Hermes Trismegistus* (2011), about the influence of Hermeticism and its mythical founder, the 'thrice greatest Hermes', on western consciousness, I touched on the idea of human beings as 'cosmic caretakers', as individuals given the responsibility of 'taking care of the cosmos' – no mean task, as I'm sure readers will agree. Although for centuries Hermes Trismegistus was believed to have been a real person who lived at 'the dawn of time', and who received a primordial 'divine revelation' – the 'perennial philosophy' that is at the heart of much of western spiritual thought – he is now thought to have been a fictional figure, devised by the authors of the Hermetic writings, who lived in Alexandria in Egypt in the first few centuries after Christ. In the *Asclepius*, one of the books making up what is known as the *Corpus Hermeticum*, the body of mystical writings on which Hermeticism is based, Hermes Trismegistus tells his student Asclepius that man is a creature of two natures, of, indeed, two worlds. Man is, according to Hermes, a creature of the natural world, of the body and the senses, and as such is subject to all the laws and limitations that come with 'living in the material world'. But he is also an inhabitant of another world, that of mind, spirit, the soul, consciousness, which, in essence, is free from the limitations of his other nature.

How this came about is told in the Hermetic creation myth.[1] In the *Poimandres*, perhaps the best known of the Hermetic books, Nous, or the Universal Mind, explains that after the creation of the world, he thought it good to create a being like himself who could enjoy his work. So he created man. For a long time, the idea that Hermetic man was created in the image of his creator suggested that the Hermetic books borrowed elements from the Judeo-Christian tradition. In that tradition, too, man is created in the image of God. Recent scholarship, however, argues that the author of the *Poimandres*, who remains unknown, came to the idea independently. Whatever the case, in two powerful spiritual traditions that have had an enormous influence on western consciousness, the same idea, that man is made in the image of the divine plays a central role.

Unlike in the Judeo-Christian tradition, but similar to the Platonic and Gnostic creation accounts, in the Hermetic account the actual work of making the cosmos was undertaken by a 'second Nous', a demiurge or 'craftsman', created by the first Nous to carry out the job. When man saw what the craftsman had forged, he marvelled at its beauty, and quite understandably, wanted to be a creator himself. Nous, his Father, loving man, agreed. The craftsman did as well, and happy to share in his work, he gave man some of his power. He gave him a share of the 'seven spheres' which encircle the earth, the seven spheres being the orbits of the seven ancient planets, Moon, Sun, Mercury, Venus, Mars, Jupiter, and Saturn. These seven spheres govern what takes place on earth; in Hermeticism, as in astrology, they are the source of our 'destiny' and 'fate'. Readers familiar with the history of astronomy will know that they are the seven planetary spheres that pre-Copernican astronomers believed encircled our earth, and they will also know that the ancient astronomers believed that the earth itself was at the centre of the cosmos.

Now, in the Hermetic creation myth, the cosmos and the earth were originally formed through the action of the creative Word, the *logos* or Mind. But by the time that man is created, the Word has left the earth and returned to its source in Nous, leaving behind a world of mere matter. It is explained in the creation account that the material which the craftsman uses to create the cosmos is a kind of 'grim darkness' that originates in the chaos which precedes creation. When the *logos* leaves it, it returns to its original state. To readers of contemporary cosmology, 'grim darkness' sounds rather like the 'dark matter' of which we are told most of the universe is made. It is still a beautiful world, and through the turning of the seven spheres, living things have emerged from the earth's waters. Man, curious about what the creator has been up to, desires to see the earth. He pierces the seven spheres and looks down upon the beautiful world, marvelling at the craftsman's handiwork. The earth, however – we can also say Nature – sees man too, and recognising the Nous within him, desires him – apparently the earth is a woman – wanting to partake again of the divine Word. Man, too, sees his reflection – and hence that of Nous – on earth's waters, and becomes enchanted with it, much as the youth Narcissus does in the Greek myth. No sooner does man wish to be with the earth then he drops from his heavenly perch through the seven spheres and enters into a form without the Word, that is, a body; in other words, matter. (Up until then he has been in a solely spiritual immaterial form.) It is through this descent from beyond the seven spheres to the earth that man becomes a creature of two worlds. Passing through the seven spheres, he absorbs their character and becomes subject to their laws. Wrapped in the arms of the earth, he is subject to all its limitations, to the constraints of mindless, spiritless matter, and the necessities inflicted on him by the dictates of 'destiny' and 'fate'. But within him still glows a spark of his origin, his birthright from the world beyond, and it is this connection with his source, with its freedom and power that will save him.

The Gnostics

In many ways this myth is very similar to the account of man's place in the cosmos given by the Gnostics, who were contemporaries of the ancient Hermeticists. The Gnostics were early Christians who embraced an interpretation of Christ's teaching very different from what became the official church. As in the Hermetic myth, for the Gnostics, creation is the work of a second Nous or God, but in their case it is an entirely disastrous affair. This second God, whom they call the demiurge, or 'half god', is something of an idiot; at least he is so inflated with his sense of power and importance that he comes to consider himself the true God. For the Gnostics, this demiurge is Jehovah, the God of the Bible – William Blake called him 'Old Nobodaddy' – and the world he has created is a kind of trap, an evil realm of falsity and oppression. Yet like the Hermeticists, the Gnostics, too, believed that a spark of the true God – the God beyond the world – was hidden within them. The aim of their spiritual practices and beliefs was to awaken this spark and release it, so that they could return to their source, beyond creation.

This notion of the world being a trap, and its creator a kind of demon, has had a powerful influence on western consciousness. Although for many centuries, the only source of information about the Gnostics were hostile accounts written about them by church fathers, who saw them as heretics – indeed the church was particularly successful in wiping them out – in the last century or so, the work of many different scholars has provided a different, broader view of these early Christians and their ideas.[2] Gnostic themes and what we can call a Gnostic sensibility have become a part of the modern mind. They can be found in the work of the psychologist C.G. Jung. The idea of life as a kind of prison which we must escape is at the heart of the 'Fourth Way' of the enigmatic esoteric teacher G.I. Gurdjieff.[3] Gnostic ideas can be found in the work of novelists like Hermann Hesse and Thomas Pynchon, in the philosopher Martin Heidegger, for whom man is 'thrown' into the world, and in less high-brow sources, such as the science fiction of Philip K. Dick and the films in *The Matrix* (1999–2003) series, and Peter Weir's film *The Truman Show* (1998), about a man who discovers that his entire life has actually been a television program. Another film, Alex Proyas' *Dark City* (1998), though less well known, is the most Hermetic of the lot, with its hero discovering that, not only is the world he lives in false, he himself is a kind of god.

It is a welcome sign that Gnostic ideas have made their way into the cultural mainstream. They lead us to question the status quo and seek the truth. But there is another side to this development. This idea of the world as false, as a kind of prison, has, I believe, led to, or at least certainly added to, our sense of uncertainty and insecurity, to our anxiety and paranoia. The kind of 'conspiracy consciousness' that permeates much of our postmodern life is a kind of Gnosticism; at least it shares in the sense that, in the words of a Bob Dylan song, *something* is happening but we don't know what it is. Powers greater than ourselves – the government, big business, aliens, the unconscious, or the 'cultural forces' invoked by much postmodern thought – control our lives, and this feeling of being manipulated adds to the general sense of helplessness which is a strong current in contemporary life. This sense of helplessness can lead to some undesirable effects, such as random violence against the 'system' or a general 'retreat from life'. It can lead to cynicism and a kind of generic nihilism that accepts that 'nothing is true', with the corollary that, everything, then, is permitted. It also encourages the kind of ironic world-weariness associated with some forms of postmodernism, the 'been there, done that, got the T-shirt' sensibility that informs much of our jaded tastes. As Colin Wilson pointed out decades ago, modern man suffers from what he calls 'the fallacy of insignificance', the sense that nothing we do really matters, that life is meaningless, and that, in the long run, 'you can't win'.[4] This is an extremely unhealthy state of mind, and if this book has a central aim, it is to show that it is wrong.

The leap

It is true that the Hermetic philosophy shares some elements with the Gnostics. We know that both groups knew of each other, as the famous 'Gnostic gospels' found in Nag Hammadi in Egypt in 1945, included some Hermetic texts.[5] But there is one central sense in which they are radically different. The Hermeticists believed, as did the Gnostics, that they had fallen from a free, limitless spiritual world into this severely limited world of necessity and constraint. But unlike the Gnostics, they did not believe that this world was a trap or that they were prisoners in it. As we have seen, for the Hermeticists, man's descent from the higher spiritual world into this world of space and time, of constraints and limitations, was not the work of an evil or idiotic demiurge, but came about through man's love of the earth, and the earth's love of man. There is even a sense in which this descent wasn't a 'fall', as it is considered in the Gnostic and Judeo-Christian traditions. Rather it was a jump, or, to take a leaf from the Danish religious philosopher Kierkegaard, a leap, of faith perhaps. Although some of the thinkers and philosophers I will look at in this book do consider that man's fall was the result of some cosmic catastrophe or crime, there is a sense, I think, in which we can see it as a freely chosen act, a willing embrace of a tremendous responsibility and obligation.

The Hermetic philosophy sees it as such. When Asclepius asks Hermes Trismegistus why man has a dual nature – one of matter and one of spirit – Hermes explains that it is so he can 'raise his sight to heaven while he takes care of the earth,' and so he can 'love those things that are below him' while he is 'beloved by the things above.' Asclepius himself, when asked about man's need for a body, explains that it is necessary so that we can take care of creation. Asclepius tells his listeners that Nous gave man a 'corporeal dwelling place' and 'mixed and blended our two natures into one', doing justice to our twofold origin, so that we can 'wonder at and adore the celestial, while caring for and managing the things on earth.'[6] For the Hermeticists, man finds himself on earth not as the result of some cosmic catastrophe or a 'fall from grace' or because he is trapped on it through the machinations of an evil idiot god, but because he has a particular mission to accomplish here. He – we – are here for a reason. As the Gnostics did, Hermetic man struggled against the constraints of the world, the snares of matter and the body, the limitations of the flesh, the prison house of the cosmos, the destiny and fate of the seven spheres. But this was not in order to escape from creation, but in order to take our rightful place within it: to embrace the obligations and responsibilities that come with being 'caretakers' of the cosmos.

In the cosmos but not of it

But if we cannot take care of the earth or the cosmos if we escape from it, neither can we take care of it if we are only a part of it, like everything else, subject to its laws, limitations, and constraints. Caretaking seems to imply some position *outside* or *above* what you are taking care of, whether it is children, a pet, or someone's flat. If I am taking care of my children, I cannot act like a child myself. Or I can only briefly, in play, and only on the condition that, when necessary, I assume full responsibility as an adult. There is a Sufi saying, which is also in the Bible, that tells us that we should be 'in the world, but not of it.' This tells us that although we cannot avoid pain, suffering, triviality, falsehood, inequity and the other evils in the world we do not have to submit to them, as difficult as that may be. In a sense we can say that in order to take care of it, we need to be *in* the cosmos – which we clearly are, at least physically – but not *of* it. The Hermetic account of man and the world seems to agree. Man is made of the stuff of the world, the 'grim darkness' that preceded creation. But he is also made of 'higher' stuff, the mind. So, at least according to Hermes Trismegistus, while we are in the world, we are not completely of it.

This idea, that we are in the cosmos but not necessarily of it, may seem strange to readers unfamiliar with the Hermetic tradition. Nevertheless it was, in different ways, shared by some important thinkers who were, more or less, within the western intellectual mainstream. In the early part of the last century, philosophers as different as the phenomenologist Max Scheler (1874–1928), the cultural philosopher Ernst Cassirer (1874–1945), and the Christian existentialist Nikolai Berdyaev (1874–1948), all came to a similar conclusion, although by very different paths. Each contributed to a movement in twentieth century philosophy known as 'philosophical anthropology', an attempt at arriving at a *metaphysical* account of man, broader and more holistic than the reigning reductive, materialist ones. In different ways Scheler, Cassirer, and Berdyaev, arrived at the same conclusion: that it is impossible to explain man adequately in terms of his place in society, his animal origins, his physical constitution, or the deterministic laws it is subject to. Each argued that man's essence is *creative*, that human consciousness brings a new dimension, a new world, into being, and that any attempt to reduce this to the laws that govern the physical world is not only doomed to failure, it results in a world empty of all meaning and value.

Cosmic amnesia

For the Gnostics, then, we are spiritual beings, trapped in an evil physical cosmos, and our only salvation lies in escape. For the Hermeticists, we can say that we are spiritual beings with a mission, but *we have forgotten it* and our salvation lies in remembering it. And, if the idea that we are caretakers of the earth and cosmos is correct, then it is not only our own salvation, but that of creation itself, which is at stake. Wrapped in the arms of Nature, we have fallen asleep and we dream that the limited, constrained world of time and space and matter, the everyday world we know so well, is the only reality. As long as we remain sunk in this dream world, this is true: it *is* the only reality. And as we are, by most official accounts – which emerge within the dream – only insignificant transitory specks in a vast, non-human universe, which has existed, we are told, for billions of years, the idea that we are in some way responsible for it, is laughable. Yet there are moments when we wake, briefly, from the dream, when some vague memory of another life, and another world, flits across our consciousness, when we somehow remember who we are and why we are here, and when the sense of some enormous mystery comes over us and disturbs our slumber. For most of us, these moments are little more than a brief, strange feeling, which, if we notice it at all, we chalk up to being 'weird' and soon forget about. Some of us, however, are troubled by them, and by the feeling of unreality they cast upon the solid, unavoidable world we bump up against each day, and by the nagging sense of *having forgotten something* that they seem to produce.

As a teenager in New Jersey in the 1970s I read a novel by Doris Lessing, *Briefing for a Descent Into Hell* (1971), which had a powerful effect on me. In it a man is found wandering around the Thames Embankment, suffering from amnesia, and raving about fantastic adventures in other worlds. He is taken to a mental ward where the doctors, trying to 'cure' him, subject him to a battery of drugs and electro-shock therapy. Throughout the book there is the sense that, although from our commonsense everyday point of view – the view of the doctors – he is quite clearly mad, from another perspective it is unclear if his remarks are simply ravings or real memories of some other existence. There is some mission he is trying to remember, some important purpose that he has forgotten, and which the doctors, with their drugs and electro-shock, only make more difficult to recall. As an *angst*-ridden teenager it was easy to identify with the hero and to see the doctors as agents of the 'establishment', trying to force him to accept the reality of a world he *has seen through*. It was only years later that I discovered that the book is considered a work of Gnostic fiction. I can also remember a science fiction story in a comic book when I was younger, about a man investigating possible aliens, living in disguise on Earth. All his leads turn out to be false except for one, and when he tracks this one down he discovers that the alien is himself. I didn't know at the time that I was reading a version of the Hermetic account of man, but as I later discovered – and wrote a book about – popular culture is often a source for disseminating ideas that mainstream 'high' culture considers nonsense.[7]

Repairing the universe

It was while writing about this human role as a cosmic caretaker that I recalled similar themes from other spiritual traditions. Some emphasise different aspects of the caretaker role, and some push that role into more active, creative areas. In these man is seen as not only a caretaker, in the passive sense of having something already complete, finished, entrusted to his care, as a custodian or curator of a museum is. He is regarded as a co-creator of the cosmos itself, an idea I explored in other books, specifically in the context of the ideas of Rudolf Steiner.[8] And in some versions he is even seen as someone responsible for correcting the mistakes God – or whoever – made when creating the universe.

In the Jewish mystical tradition of Kabbalah, for example, there is the idea of *tikkun*, which is generally translated as 'repair'. In the Judaic tradition, as in the Christian, God is usually seen as perfect, omnipotent, and infallible, but in the tradition of Lurianic Kabbalah, stemming from the teachings of the great sixteenth century Kabbalist Isaac Luria (1534–1572), this isn't the case. According to Luria, when God created the world, something went wrong, and He created man in order to correct his mistakes, to repair the damage caused by his blunder. This surely gives man an exalted position, but some Kabbalistic interpretations go even further, and suggest that God made his cosmic mistakes *on purpose*, but unconsciously, so that he would have to create man in order to complete the work of creation. In this sense, God suffered from a kind of Freudian slip, rather like when we leave our umbrella at a house we would like to visit again, but aren't consciously aware that we do. In this interpretation, God has a 'dark side', unknown to himself, and the fractures and cracks that run through creation were planned by his unconscious, so that He would have to call in man to do the repairs. The inference is that the work of creation cannot be completed without our contribution, and some thinkers consider this to be so essential that, in the words of the Cretan writer and poet Nikos Kazantzakis (1883–1957), through it we become the 'saviours of God'.

This idea of man as a 'repairer' is also at the heart of the work of the French eighteenth century mystical philosopher Louis Claude de Saint-Martin (1743–1803), who during his life was known as the 'Unknown Philosopher', a pseudonym he used for his writings. For Saint-Martin, 'The function of man differs from that of other physical beings, for it is the reparation of the disorders in the universe'. Saint-Martin's vision is within the context of Christian mysticism, but he shares with the tradition of *tikkun* and Hermeticism the idea that man has a crucial role to play in the work of creation. Indeed, for Saint-Martin, man is, in a very real sense, the entire purpose of the universe, the answer to its mystery, the key to its existence. In some ways Saint-Martin, and other mystical thinkers with similar views, like the eighteenth century Swedish scientist and religious thinker Emanuel Swedenborg (1688–1772), seem to anticipate some contemporary scientific ideas, which argue that the universe itself is designed in order to produce intelligent life like ourselves, what is known as the 'anthropic cosmological principle'.

The fallacy of insignificance

Yet, even during his time, more than two centuries ago, Saint-Martin recognised that man suffered from a sense of insignificance. In fact, I first came across Saint-Martin's ideas in the book by Colin Wilson in which he analyses 'the fallacy of insignificance,' *The Stature of Man* (1959), a study of the 'loss of the hero' in modern literature.[9] At the beginning of the book, Wilson quotes Saint-Martin. Men, Saint-Martin writes:

> ... have believed themselves to be obeying the dictates of humility when they have denied that the earth and all that the universe contains exists only on man's account, on the ground that the admission of such an idea would be only conceit. But they have not been afraid of the laziness and cowardice which are the inevitable results of this affected modesty. The present-day avoidance of the belief that we are the highest in the universe is the reason that we have not the courage to work in order to justify that title, that the duties springing from it seem too laborious, and that we would rather abdicate our position and our rights than realise them in all their consequences.

'Where,' Saint-Martin asks, 'is the pilot that will guide us between these hidden reefs of conceit and false humility?' Where indeed? Trying to chart a course between this Scylla and Charybdis is one aim of this book. But if it was true in Saint-Martin's time that man has avoided the belief that he is 'the highest in the universe', it is certainly even more true today. Today any suggestion that we are in any way 'special', that we are significant or somehow central to the universe – let alone that it exists on our account – would be met with sarcastic laughter or self-righteous indignation, depending on who you were talking to. Yet it isn't only obscure mystical philosophers who believe we sell ourselves short. One reader of Wilson's *The Stature of Man* was the humanist psychologist Abraham Maslow (1908–1970), best known for his concept of the 'peak experience', sudden moments of almost mystical delight that, Maslow argues, come to most psychologically healthy people. Another of Maslow's ideas that chimed well with Wilson's concern over the loss of the hero in modern consciousness, was what he called the 'Jonah complex', after the Biblical prophet who tried to avoid the responsibility God placed on him. Maslow asked his students if they expected to do something important, to excel at their work, to make a significant contribution to psychology or society. All were diffident and none raised their hand. Maslow looked at them and said, "Well, if not you, then who?" Maslow saw that we invariably feel that *someone else* will be successful, creative, outstanding, accomplished, but that to expect that of ourselves is a kind of egotism, or a foolish overestimation of our abilities, certainly in bad taste.

Yet this modesty is of the same false character that Saint-Martin recognised in his contemporaries, and is really a defence against living up to the demands made on us by our higher, better selves. Maslow recognised that although the fear of failure is common and understandable, we also suffer from a 'fear of success', a fear of living up to our full potential, of the responsibilities and obligations this entails, as well as of the ostracising our less exceptional fellows will direct at us. Like Jonah, we want to avoid any responsibility that will set us apart from the average. We reject it, and want to remain an anonymous member of the herd. Yet such sheepishness is just as much a neurosis as any other, and by embracing it we are, according to Maslow, displaying a kind of psychological illness, and blocking our way to 'self-actualisation', Maslow's term for the process of becoming what he calls 'fully human'.

Fully human or only human?

But as Saint-Martin recognised two centuries earlier, being 'fully human' is something most of us avoid. In our climate of insignificance, we are more comfortable with the 'only human', with associating our humanity with weakness, sickness, mediocrity, and the collection of appetites Maslow calls 'deficiency needs', our hunger for the three S's: security, sex, and self-esteem. Being 'fully human' makes demands on us, it is a kind of existential *noblesse oblige*, which requires that we apply high standards and aims and values to ourselves and our actions. If we are 'only human', as many of us prefer, then not much can be expected of us. We are let off the hook, can let it 'all hang out', and can get by, as the cliché goes, as a 'good enough' human. But good enough for what? If Maslow and Saint-Martin are right, then certainly not good enough to take on the responsibilities that being fully human demand.

Trying to meet the demands of being 'fully human', we encounter difficulties, not only from our own reluctance and fear, but also from the ideological atmosphere in which we live, from the *Zeitgeist*. For one thing, science tells us that the universe is too large and too old for us to be of any significance in it. The philosophy that emerges from this belief has some interesting adherents, among who are the horror fiction writer H.P. Lovecraft (1890–1937) and the contemporary social philosopher John Gray. Science also tells us that the universe is meaningless, a pointless product of an accidental explosion, and that eventually any life within it will be extinguished and it itself will dissipate into an endless cosmic emptiness. If that isn't bad enough, science also tells us that we ourselves are the product of an equally meaningless process, and that far from being the answer to the riddle of the universe, there was absolutely no intention in our being here at all.

Slime moulds and giraffes

Scientists are not the only ones who reject the idea that humanity is in some way special. In recent years, concerns over our ecological crises, our rampant abuse of natural resources, global warming, and other environmental problems have led some well-meaning people to suggest that rather than a cosmic caretaker, man is really a blight on the planet. To them the Biblical injunction that man has dominion over the earth, has given us carte blanche for the selfish exploitation of nature. Paradoxically, the science that tells us that we are merely meaningless accidents in an accidental universe, has also produced the technology that has allowed us to fulfil that Biblical injunction, and made us masters of the world. Ironically, in our secular age, man is at once reduced to a nullity and made lord of creation. Yet it is this mastery that many nature-orientated people argue is in the process of killing Mother Earth, and has made us the most dangerous animal alive. Quite rightly, they say we must concentrate on saving the planet, and the best way to do this – the only way some insist – is for man to 'get back to nature', to recognise that he is no better or more important than a slime mould or a giraffe, and that any idea that he is, is precisely the source of our problems.

While I have nothing against slime moulds and giraffes, this 'biocentric' view has its own problems. And if Hermes Trismegistus, Kabbalah, Saint-Martin, and others are right, it is in fact dangerous.

According to them, the only way we can save the planet – the cosmos, in fact – is by recognising that we are something *more* than nature, something *more* than animals, and by taking the responsibilities that come with this seriously. (If nothing else, the recognition that no animal wants to save the planet should give us pause for thought.) By abandoning our humanness and embracing our animal roots – which, no matter how hard we try, we can never feel completely comfortable with – we are giving up our one possibility of saving anything. What that 'something more' is that sets us apart from the rest of creation, and what its relation to the cosmos may be, is something I hope to discover in this book. And strangely, one of the curious things I discovered while gathering my material, is that in many essentials, science itself shares the Hermetic vision of man as the answer to the riddle of existence, as a co-creator of the cosmos, although this more optimistic vision is usually obscured by the more pessimistic view of ourselves as insignificant creatures in a pointless, purposeless universe.

Whistling in the dark?

As mentioned, the idea of us having any significance in the universe is bold, perhaps too bold for many of us to swallow. I have to admit that more than once while writing this book I found myself thinking 'Oh come on, isn't this a bit too much?' and wanting to push the whole idea aside. For some readers the idea may seem merely one more expression of our inveterate self-importance, a trumpeting of our dubious standing as the most dominant species on the planet, and a clatter of applause at our triumphs over the rest of creation, if it is not merely a cosmic whistling in the dark. Let me assure readers that nothing could be further from my intention, and that I share with them their disdain for such boorish self-congratulation. Although I do think we suffer from a kind of cosmic low self-esteem, and have a poor self-image, my aim here is not to rack up reasons to feel pleased with ourselves, or to encourage a session of mutual back patting. I certainly am not interested in hoisting banners celebrating our eminence, although I do agree with Maslow, Wilson, and other writers and thinkers I will look at, that we have come to see ourselves in a false light, one which encourages countless *mea culpa*'s (in our idiom, 'my bad') and discourages any feeling of self-confidence and assurance. My aim is to try to steer a course between the conceit which is often an excuse for complacency, and the false modesty that Saint-Martin argues is really an avoidance of our obligations. And having found a way through these hidden reefs, on which more than one seafarer has found himself wrecked, I would like to discover what lies ahead.

Outline of the book

In what follows I will try to understand what is involved in the idea of ourselves as caretakers of the cosmos, what it can mean in our lives and what, if anything, we can do about it. My starting point is the Hermetic vision of ourselves as belonging to two different worlds, of man as a microcosm, or 'little universe'. These are longstanding ideas in our esoteric 'counter tradition', yet according to recent developments in neuroscience, they may actually be rooted in the very structure of our brains. It may be that the 'other world' we have 'fallen' from, have brief haunting memories of, and a yearning nostalgia for, is *another mode of consciousness* that is our birthright, but which, through our evolutionary development, we have come to ignore at, it may be, our peril. From this vantage point I will look at how our caretaking can apply to our personal lives, to our relation to society, and to nature. Given the uncertain times we live in, it is understandable that this aspect of repairing the cosmos may strike most of us as of central importance. Indeed it is. Speculations on our place and significance in the cosmos and other modes of consciousness can be thrilling and highly entertaining, but if they don't lead to actual changes in how we live, they are merely pleasant daydreams and can even make us less capable of dealing with reality than we already are. To paraphrase Hermes Trismegistus, we can have the stars in our eyes, but we must have our feet on the ground. Swedenborg, one of the most grounded of spiritual thinkers, entreats us in our daily lives to 'do the good that we know', and it is from that humble beginning that our cosmic caretaking should begin. Applying Maslow's ideas about our 'hierarchy of needs' to society, I will look at what evidence there is to suggest that we may be moving beyond the need for self-esteem into more creative levels, those of the 'fully human', and how this relates to Max Scheler's 'hierarchy of values'. And through the ideas of Gustav Fechner, Goethe, and others, I will explore how we can better understanding our relation to a living nature, without losing our independence as creative agents within it.

But while what we might call our 'hands on' form of caretaking is crucial, and while we all need to recognise the preciousness of the world which has been entrusted to our care, there is another form of our responsibility of which we also need to be mindful. And being 'mind full' of it is indeed its core.

As I do in my book *A Secret History of Consciousness* (2003), in the later chapters of this book I will explore the ways in which our own minds are involved in actually creating the world we experience and subsequently care for. From a variety of different perspectives – quantum physics, neuroscience, phenomenology, the philosophy of language – it is becoming more and more clear that the universe we live in is a 'participatory' one, in which mind and matter, the inner world and the outer one, are not, as our commonsense view suggests, radically different and opposed realities, closed off from each other, but are different aspects of a single shared reality. It seems increasingly clear that the barriers between these two worlds are not as impermeable as we have believed. Our inner worlds, it seems, are not isolated islands of consciousness, floating on the surface of a dead, material world that is oblivious of them, and on which they have no effect. In some strange, still inexplicable way, our inner worlds *participate* in the world outside us, something less modern, more 'primitive' people still experience, but which to us seems fantastic nonsense. Synchronicities, those strange meaningful coincidences, in which some thought or feeling in our inner world is paralleled by an event in the outer one, and other paranormal experiences, are one way in which this participation manifests, but there are others. One idea that runs throughout this book, as it does in my others, is that at an earlier stage in our evolution, human consciousness was much more 'embedded' in nature, as animals are today, and that we did not experience then, as we do now, separate outer and inner worlds, but a free flowing movement between the two. Indeed, the separation from nature that we experience now can be seen as a result of our 'fall', not from a heavenly paradise, but from our evolutionary past. But if this fall was really a leap, it is my belief that *nature itself* pushed us out of her warm embrace, as bird pushes her chicks out of the nest, in order to get them to fly. At some point in our evolutionary past, consciousness became aware of itself and the world as two different things, and we can say that at that point, humanity and the cosmos itself, 'began'. As Marie-Louise Von Franz, one of Jung's most important disciples and an original thinker in her own right, suggests, most creation myths are really about the rise of consciousness out of an unconscious, undifferentiated ground. If that is the case, then 'getting back to nature' is the most unnatural thing we can do, as it would be moving in the opposite direction from nature's own intentions.

I believe that nature, the world, the cosmos, separated us off from itself in order for it to become *conscious of itself through us*. It is in this way, through our own increasing consciousness, that the work of creation is completed, or at least carried on. Drawing on the work of different 'participatory' thinkers, it is my belief that our evolutionary task now is to regain an experience of participation and all that it entails, without losing our independence as conscious egos, capable of free will and creative action, something our ancestors, more at one with the cosmos, lacked. Our task, then, is to become *more* conscious, not less, which means facing the sense of separation from the world firmly, and *getting through it*.

Isolated, alienated, feeling adrift in the cosmos, it is understandable that we would want to return to the nest, to press my mother bird metaphor a bit further. But as pleasant and blissful as this might be – and consciousness has devised many delightful ways in which to free itself of the burden of itself – it would really be a shirking of our business as caretakers. We cannot return to an earlier stage of our evolution, just as we cannot become children again, try as we may. Nor can we stay as we are. In a living universe, which I believe ours is, stagnation is just another word for death. We must press forward into what the poet Walt Whitman called 'the unknown region'. It is here that our caretaking adopts a more adventurous character, and in the last section of the book I will explore what this might entail. Here we are no longer taking care of a cosmos already made nor repairing the cracks and fissures left by an inadequate craftsman, but are bringing *new* worlds into being. In the last chapter I will look at what this might mean, and how the ephemeral flicker of consciousness – at least as seen from the point of view of materialist science – housed in the insignificant creatures of a transitory planet, may actually be the key to saving the cosmos itself from oblivion. A fate that, at least according to the latest scientific accounts, is inevitable.

Save the planet? Yes, assuredly. But why stop there? Why not save the universe while we're at it?

1. The Other Side

In the beginning was the spill, at least according to the account of creation given by Isaac Luria, the great Kabbalist of Safed, in the sixteenth century in what was then Ottoman controlled Palestine. What exactly I mean by 'spill' will be explained shortly. Kabbalah is a Hebrew word meaning 'receiving' and is the name given to the mystical tradition within Judaism. Kabbalah claims to explain the nature of the universe and man's place within it, just as the Hermetic writings do, and is said to be the esoteric or inner teaching hidden within the exoteric, or outer reading of the Torah. Perhaps its most well known feature is the *Otz Chiim*, or Tree of Life, a symbolic mapping of the divine energies at work in the world, with its ten *sephiroth*, or 'vessels', which has become almost as ubiquitous in popular occultism as the Tarot, with which it is often associated; erroneously, it seems, as most Judaic scholars argue. At Kabbalah's heart is the relation between creation, the finite, physical cosmos, and its infinite, unmanifest source, called the *Ein-Sof*, which means 'limitless' or 'unending'. This is a sphere or dimension of what we can call 'negative existence', which really means that it is a plane of reality that our finite human minds are incapable of comprehending, and not merely a simple emptiness. The *Ein-Sof* is so 'other' than what we normally perceive as reality, that we cannot make positive statements about it. Any positive statement about it, would, by definition, be a limit on it, and as it is limitless, cannot apply.

This negative existence has parallels in other spiritual traditions. It is the Neti-Neti ('not this, not that') of Hinduism, the *sunyatta* or 'void' of Buddhism, the Pleroma of the Gnostics. It is even part of the Christian faith. The Athanasian Creed, in use in Western Christianity since the sixth century, declares that the Father, Son, and Holy Spirit are 'uncreated' and 'unlimited', and in more than one place in scripture we are told that God 'has neither body nor parts'. It is at the heart of the negative theology of Meister Eckhart, forms the *Ungrund* of Jacob Boehme's difficult alchemical writings, and can be found in the *Nichts* or 'positive nothingness' of Martin Heidegger's 'fundamental ontology' and his predecessor Hegel's tortuous dialectic. In more scientific terms, a similar idea, but without the spiritual connotations, seems to be present in the way scientists talk about the state of things 'before' the Big Bang. I put 'before' in quotation marks because, according to most accounts, there was no before before the big bang, a confusing situation, to be sure. A kind of non-manifest ground of our everyday reality also seems to be involved in the 'implicate order' of the physicist David Bohm. There are other expressions of the idea, but I think this will suffice for now.

Just as the Hermetic teachings are supposed to have been revealed to Hermes Trismegistus at the dawn of time, Kabbalah is said to have been revealed to Adam 'in the beginning' in Eden, but as with the *Corpus Hermeticum*, a more wieldy dating for its conception is available. Historically, Kabbalah comes to light in the twelfth and thirteen centuries in Spain and Southern France, emerging from earlier forms of Jewish esotericism, such as the *Merkabah* or 'throne' or 'chariot' mysticism, that centres on the vision of a mystical chariot in Ezekiel. The *Zohar*, one of the most important texts in Kabbalah, was said to have been discovered in 1286 by Moses de León, a thirteenth century Spanish Kabbalist. León claimed that it was written in second century Israel by the miracle-working Rabbi Shimon Bar Yohai, but most modern scholars believe it was written by León himself. Although mainstream Judaism considers Kabbalah nonsense, it is at the centre of Hasidic Judaism, the more ecstatic form of the faith.

Isaac Luria

Isaac Luria is considered the 'father of contemporary Kabbalah', and it his interpretation of the mystical tradition that concerns us here. A story has it that one day, Isaac's father, an Ashkenazi Jew, was in the synagogue in Jerusalem (where Isaac was born) studying, when he was visited by the prophet Elijah. He told him that he would have a son who would deliver Israel from the *klipoth* ('shells' or forces of evil) and transmit to the faithful their *tikkun* ('reparation'), and that he would make clear many of the mysteries of the Torah and explain the secrets of the *Zohar*. The story is most likely apocryphal, but it sums up Luria's short life admirably.

By the age of twenty-two, Luria had become devoted to the study of the *Zohar*, which had only recently been printed for the first time. He became a hermit, living for seven years on the banks of the Nile in a small cottage. During this time he saw his family (he was by then married) only on the Sabbath, and had taken a vow of silence, speaking only when necessary, even to his wife. Legend has it that during his solitude he was visited on more than one occasion by Elijah, who tutored him in the mysteries.

In Safed in 1570, Luria became a student of the Kabbalist Moses Cordovero (1522–1570), author of *Pardes Rimonin*, or 'Garden of Pomegranates', an attempt at a systematic presentation of Kabbalistic philosophy, linking it to previous teachings, and trying to bring different variants together into a coherent whole. Considered a classic and standard text, its presentation was only superseded by Luria's own ideas. Luria's tutelage under Cordovero lasted only a short while, a few months at best, and when Cordovero died, the circle of students looked for new guidance. Their search didn't take them very far, and Luria accepted the role as teacher. Elijah's prophecy, apocryphal or not, seemed to be coming true, and for the remaining two years of his life, Luria laid down the groundwork of a new reading of the revelations.

Luria himself did not write much. His literary remains consist of a few poems, but after his death his ideas were disseminated by his disciple Chayyim Vital (1543–1620), who played a kind of Plato to Luria's Socrates, making notes of his lectures and compiling these into books, such as the eight volumes of his *Otz Chiim*. It was through Vital's efforts that Luria's vision achieved the prominence it has.

Tzimtzum

Luria's re-visioning of Kabbalah came about because of problems in answering certain basic questions, such as how God or the *Ein-Sof*, which is infinite, could create a finite world like our own. (A similar and related conundrum is how God, who is all good, could create a world like ours, that contains evil.) Luria's answer was ingenious. Luria came up with the concept of *tzimtzum*, which means 'concealment' and 'contraction'.[1] According to Luria, in order to create our world – the universe – God had to 'withdraw' or 'contract' a part of his infinite being, to create, as it were, a 'hole' in himself within which a void or empty space could exist. We can, then, think of our entire universe as a kind of hole in God. And as our universe is, by our standards, fairly large – just how large we will soon see – we can understand why the Kabbalists believed that trying to grasp the reality of the *Ein-Sof* might be beyond our powers. Once the *tzimtzum* created the void, Adam Kadmon, the Primordial Man, appeared. This idea, of an archetypal human form, which is itself the universe, can be found in other spiritual traditions. Swedenborg, for example, speaks of the Grand Man, an idea that was carried on by his erstwhile disciple William Blake (1757–1827), who speaks of the 'Human Form Divine'. The Hindu Vedas speak of Purusha, or 'Cosmic Man', from whose body the world is created. The basic idea is that the universe itself is ultimately in the shape or form of man, a very early form of what today we call the anthropic cosmological principle, the idea that, as mentioned earlier, the universe is designed in such a way that it will bring about the creation of intelligent life.

Out of the eyes, nose, mouth, and ears of Adam Kadmon come flashing lights, emanations of the divine creative energies. These form the *sephiroth*, or vessels, designed to contain these energies, as well as the twenty-two letters of the Hebrew alphabet, the holy *otiyot yesod* or spiritual 'building blocks' of creation. (Much as the Greek philosopher-mystic Pythagoras believed that the world was made, literally, of numbers, the Kabbalists believed it was made out of the Hebrew alphabet.) As in many spiritual traditions, in Kabbalah words and language have mystical powers, and this is why the true name of God is never to be pronounced; instead it is referred to as the Tetragrammaton, the YHVH or Jehovah of convention. Were God's true name spoken, it would, Kabbalah tells us, destroy the world.

When the vessels break

The vessels designed to hold the divine energies and archetypal forms with which the *Ein-Sof* will create the world, proved unable to contain them. Just as an electrical transformer can hold only so strong a current, the *sephiroth* could not withstand the terrific power coursing through them. What happened then is known as the *shevirat-ha-kelim*, the 'breaking of the vessels' (*kelim* is another word for vessel). In this cosmic catastrophe, the vessels shatter, and the divine energies are scattered throughout the void. In one version, the divine energies are seen as waters flowing from heaven. The vessels aren't deep enough to contain the waters, and so they spill out into the emptiness, creating literally a 'holy mess'. Hence the opening sentence of this chapter. The sacred letters of the alphabet are also jumbled into a confusing nonsense. A tear appears in the seamless void and the unified divine forces separate into their opposites, male and female, causing a fracture in God and the Primordial Man. In the words of Gershom Scholem, the most respected Judaic scholar of the last century, following the *shevirat-ha-kelim*, 'Nothing remains in its proper place' and 'everything is somewhere else'. That somewhere else is our world.

As shards from the broken vessels fall, they capture sparks of the divine light, and these sparks sink into layers of darkness, much like the 'grim darkness' of the Hermetic account. Tumbling into *Sitra Achra*, the 'Other Side' – the universe as we know it – the shards come together to form husks, or shells, known as *klipoth*, a kind of negative version of the *sephiroth*, something like a Kabbalistic anti-matter (their designations as husks and shells suggest their empty character). The result of the breaking of the vessels is that the world, which the *Ein-Sof* had originally planned to be formed of the highest values – beauty, love, mercy, wisdom, knowledge – is now corrupted by their evil counterparts. And we too, who are fragments of the androgynous Primordial Man, are infected with the corruption. The *klipoth* that make up the universe are in us too, and we find ourselves here, separated into opposites, male and female, stranded on the Other Side. Hence our world is one of pain, suffering, falsehood, conflict, and the other evils we are all too familiar with.

Tikkun

This depressing state of affairs is reminiscent of the Gnostic view of ourselves as trapped in an evil world devised by an idiot god. But, as in the Hermetic account, all is not lost. Luria's dramatic version of the creation story and its subsequent cosmic catastrophe would be a paranoid's dream – or nightmare – if it were not for the possibility that the situation can be saved. In this man plays the central role. If in the Hermetic creation myth we are caretakers of the world, for Lurianic Kabbalah, we are its cosmic repairmen, here to clean up the mess caused by the spilling of the divine energies. Although the *klipoth* are within us, as in the Hermetic account, we also carry within us sparks of the divine energies. In fact, everything in the world has within it some heavenly spark. Mankind's job is to free the sparks (*netzotzim*) from the shards, through what Luria called *tikkun*, or 'repair'. Through this we will restore the shattered *sephiroth*, heal the rift between the opposites, and unify the polarised masculine and feminine aspects of God.

The German-Jewish cultural philosopher Walter Benjamin (1892–1940) combined a deep interest in Kabbalah with a commitment to Marxism. In one of his most well known essays, Benjamin wrote of what he called 'the angel of history'.[2] Where we perceive the past as a chain of separate events, the angel, Benjamin tells us, sees it as 'one single catastrophe which keeps piling wreckage upon wreckage'. This 'single catastrophe' is the result of the breaking of the vessels, and it takes place throughout history, here, on the Other Side. Benjamin had hopes that a kind of Marxist redemption would save the world, and 'make whole what has been smashed'. Yet ultimately he despaired and committed suicide while trying to escape the Nazis.[3]

Luria, and the Kabbalists who followed him, believed that we can stop the catastrophe and make whole what has been smashed. But he didn't think, as Benjamin did, in terms of some Messianic event – a revolution or even divine intervention – that would finally stop the wreckage from piling up.[4] His approach was more individual. He believed that in our lives we encounter those trapped sparks, whether in people, nature, or inanimate things, which we are uniquely qualified to liberate, just as we encounter people and situations that can help liberate our own 'sparks'. By doing this, we 'repair' the universe. But equally so, our failure to perform *tikkun* only increases the world's confusion. This being so, mankind has a crucial role to play in the world. As Sanford L. Drob, a contemporary Kabbalist, writes: 'the restoration and repair of the broken vessels is largely in the hands of humankind.'[5] As mentioned, for some Kabbalists, God or the *Ein-Sof* shattered the vessels on purpose, precisely so that man could repair them. But in this version, our work is more than repair. 'In freeing the divine sparks from the *klipoth*,' Drob writes, 'and restoring them to God, and re-establishing the flow of masculine and feminine divine energies, man acts as a party in the creation and redemption of the world, and is actually said to complete God himself.'[6] Not only are we entrusted with the responsibility of saving the universe, but, as mentioned earlier, God too.

It's a big world after all

We will return to *tikkun* and how we can clean up the cosmic mess we find ourselves in. Right now, let's take a closer look at that hole in God we call the universe. What is it like on the Other Side?

At the risk of pummelling the reader into submission with numbers, let me relate some facts and figures about the place we call home. According to the latest estimate, the observable universe seems to be a sphere of about 93.2 billion light years in diameter; that gives it a radius of roughly 46.6 billion light years. A light year, we know, is the distance that a beam of light can travel in a year. Light travels at 186,000 miles per second, so a light year equals roughly six trillion miles, or just short of ten trillion kilometres. A trillion is a 1 followed by twelve zeros. One light year is a 6 followed by twelve zeros. Six trillion times ninety-three billion would give us the diameter of the observable universe in miles, or roughly 558 with twenty-two zeros following it.[7] I say the 'observable' universe because, although the portion of the universe that we can actually observe, through the Hubble and other powerful telescopes, is large, according to the latest estimates it is only a fraction of the entire universe. That's assuming there is an 'entire' universe, meaning a single, definite object that we could in some way see as a whole, a proposition that itself entails dizzying logical problems.[8] We do not know if the rest of the universe, which we can't observe, ends in some way, or if it simply carries on into infinity, but either proposition is, like the *Ein-Sof* itself, unthinkable for us.

Exactly why there is an 'observable' and 'unobservable' universe is complex, but for our purposes we can say that there is reason to believe that there are objects in the universe so distant from us and travelling away from us at such a speed – our universe, we know, is expanding – that their light will never reach us. It is not a question of our instruments not being powerful enough. The restrictions of time and space themselves and the conditions set by the proposed Big Bang, make it impossible for us to ever detect these objects. One estimate of the size of the total universe, provided by the 'cosmic inflation' theory of Alan Guth, gives us a figure 10/23 times larger than the observable universe. I will not burden the reader with the number of zeros this gives us. The 'cosmological principle', which goes back to Isaac Newton, tells us that there is no privileged position within the universe from which to observe it, meaning that from any point in the universe, the universe looks pretty much the same, in whatever direction an observer may observe it. Given that the notion of an 'observable' universe is true from any point within it – it is not a measurement of distance itself, but of how much of the universe can be observed from any point within it – and the much larger total universe that lies beyond the 'observability barrier', it seems that the old definition of God as 'a circle whose centre is everywhere and whose circumference is nowhere' takes on new meaning. Each of us is a centre of the observable universe, and were we to travel half the distance to its edge, we would still be a centre of the observable universe. Only the parts of the universe we could observe would have changed.

Estimates for the number of stars within the observable universe vary from 3 to 100 x 10/22, or thirty sextillion to one septillion, with an estimated eighty to two hundred billion galaxies. This, of course, is only a fraction of the number of stars and galaxies there may be in the total universe. Within the observable universe, of the largest objects so far detected we find what is called the Great Wall, a sheet or chain of galaxies, known as a 'galactic filament', more than five hundred million light years long and two hundred million wide. An even larger structure is the Sloan Great Wall, so named because it was discovered by J. Richard Gott III and Mario Juric in 2003 using data from the Sloan Digital Sky Survey. It is 1.37 billion light years long, more than twice the size of the Great Wall. Another large structure is known by the curious name of Newfound Blob, a kind of cloud of galaxies, super-clusters (groups of smaller galaxies) and gigantic gas bubbles that measures some two hundred million light years in length. There is also what is known as the Great Attractor, another large object, some 150 million light years in size, that is pulling our galaxy and the rest of our 'local group' – which includes Andromeda, the galaxy closest to the Milky Way – toward the constellation Leo at a speed of four hundred miles per hour.[9] Beyond these no larger structures have yet been discovered, giving rise to what is called 'the end of greatness', a phrase that oddly parallels the 'fallacy of insignificance' with which this book is concerned.

At the opposite end from these colossal objects are the 'voids', stretches of the universe which seem empty of any galaxies, like gigantic cosmic vacant lots. Large voids are called 'supervoids', and galaxies that find themselves inhabiting a void are called 'void galaxies'. They are the hermits of the universe, galactic recluses living lonely existences in what seems sheer empty space. Most galaxies are apparently more gregarious, existing as part of a cluster, filament, wall, or sheet. I say the void galaxies exist in empty space but apparently this is not exactly true, as what appears to us as empty space is, according to the latest findings, really full of what is called 'dark matter', mentioned earlier in relation to the 'grim darkness' of the Hermetic creation myth. It is unclear exactly what dark matter is, but the universe seems full of it. In fact all of the matter and energy in the known, observable universe only accounts for 5% of the mass of the cosmos; the other 95% is missing. Dark matter, and its counterpart, 'dark energy', as well as the equally mysterious 'power of nothing', seem to confirm what mystics and occultists have been saying for centuries: that the universe, observable or otherwise, teems with unseen ('occult') forces.[10]

Big Bangs of nothing

And if the universe is big, it is also old. 13.75 billion years at the latest estimate, although some suggest 15 billion. It began in what is known as the Big Bang, a kind of Holy Grail of modern cosmology, although to some physicists the theory is not entirely unassailable. Unshakeable or not, it is curious that in at least three different creation myths – the scientific, the Hindu, and the Christian, and there may be more – *sound* plays a central role. In the Gospel of John we are told that 'In the beginning was the Word.' For Hindus, in the beginning was Om, the sacred, mystic syllable. But leave it to us moderns to start off with a bang. The vessels did shatter, after all.

Although current theory takes us to 10/-37 seconds into it – an unimaginably brief stretch of time, which, we're told, like space didn't exist until the Big Bang – we still don't know exactly what triggered the bang. Although many scientists like to say that we are closing in on the 'secret of the universe' – witness the recent excitement about the mysterious Higgs-Boson particle – any theories about *why* the bang happened at all remain speculative.[11] If we ask obvious questions such as what was before the Big Bang, scientists shake their heads and explain that we can't ask that because, as mentioned earlier, according to them, there was no 'before'. There was also no 'where' for the Big Bang to happen, as both time and space 'began' with it. Yet, as the literary philosopher George Steiner argues, 'It is intellectually, possibly morally dishonest to rule, by magisterial *fiat*, that any question as to time prior to the Big Bang is illegitimate nonsense.'[12] It is doubtful, however, if many scientists consider this.

The physicist Stephen Hawking tells us that our universe came into being through a 'quantum fluctuation in a pre-existent vacuum', which, for non-physicists like myself, can perhaps best be understood as a kind of ripple in nothing. As the world's media announced a few years back, there was no need for God to create the world, as in his book *The Grand Design* (2010) Hawking had made clear that 'spontaneous creation' was all that was necessary to get things going.[13] The universe had apparently created itself, truly a *creatio ex nihilo*. Hawking's earlier goal of coming up with a 'theory of everything' that would tie existence up into a neat bundle and leave no unanswered questions – thus making us privy to the 'mind of God' – had, it seemed, now been trumped by a 'theory of nothing.' Nothing had started the universe, he assures us, and there seems to be nothing we can do about it. It is perhaps from conclusions like these that the astrophysicist Steven Weinberg arrived at his famous pronouncement that 'the more the universe seems comprehensible the more it also seems pointless.'[14]

But once that nothing got going, boy was it something. Beginning in an infinitesimally small, extremely hot, incredibly dense state, it – whatever it was – burst into the mother of all explosions, literally. If the enormous size of the universe and of some of the objects in it, as well as the number of them, seems overwhelming, there is equal mystery in the other direction, something the sixteenth century French mathematician and religious thinker Blaise Pascal (1623–1662) recognised when he remarked that man finds himself between two infinities, the infinitely large and the infinitely small. Is it possible for us to in any way appreciate 10/-37 seconds? In between the time you read that question and consider it, the universe, according to the 'inflationary theory', had 'inflated' – like a character's head in a cartoon – by a factor of 10/50 (possibly 10/78) from less than nothing to a very big something. When this inflation stopped, the fledgling universe then consisted of quark-gluon plasma, or a 'quark soup', quarks being perhaps the most fundamental element in subatomic particles, their odd name borrowed from James Joyce's *Finnegan's Wake* ('Three quarks for Muster Mark!') by the physicist Murray Gell-Mann, their discoverer. Gluons are particles that hold the quarks together, and plasma is sometimes called 'the fourth state of matter', to differentiate it from solids, liquids, and gasses. A plasma is a gas at such a high temperature that outer electrons are stripped from atoms, producing clouds of positively charged nuclei and negatively charged electrons. Although plasmas exist on earth only under laboratory conditions, most of the matter in the universe – in the atmosphere of stars and in interstellar space – is in a plasma state, except, of course, for 'dark matter' (we don't know what state it is in). It is a curious feature of quantum physics and modern cosmology that many scientists within the fields give new entities rather whimsical names – 'quarks', 'Newfound blob' – a somewhat 'geeky' habit that accounts for the notion of a 'theory of everything' that would fit on a T-shirt.[15]

After some 380,000 years, the exploding incandescent universe cooled. For some still unknown reason more matter than anti-matter was created, a glitch in Big Bang theory which actually suggests that equal amounts of matter and anti-matter should have been created. What we know as the matter in the universe is really the extra matter left over, after the original matter and anti-matter particles annihilated each other. Were it not for this excess matter, nothing would exist. In Kabbalistic terms, we can say that there were more kernels than husks left over after the vessels shattered. As the universe continued to cool, it produced particles, protons, neutrons, and electrons that fused to form the earliest elements, hydrogen and helium, with lithium running a close third. Clouds of these gases condensed through gravity to form our stars and galaxies, and something known as 'symmetry breaking' put the fundamental forces of physics in place. Elementary particles reached their present form, with heavier elements being created within stars and supernovae, the spectacular 'death' of a star, when it literally burns itself out in a terrific explosion.

Give life a chance

According to the latest estimates, our earth itself formed some 4.5 billion years ago, roughly ten billion years after the Big Bang, from cosmic dust and gas left over from the sun's formation. It is believed life appeared on earth within a billion years after our planet formed. The standard account of the 'birth of life' suggests that self-replicating molecules accidentally emerged from the primordial soup some 3.5 billion years ago, and through an equally accidental process, over millions of years eventually turned into myself writing these words and you reading them – with, of course, quite a few different organisms in between. As with the Big Bang, the emergence of life is another example of the 'something from nothing for no reason' scenario popular with many scientists today. According to the same scenario, the consciousness I am exhibiting in writing these words – humble, indeed – and which you are employing in reading them, also emerged purely through accident, as an epiphenomenon of purely physical interactions of our brains' neurons, which are themselves the result of the purely mechanical process of evolution, the Darwinian version. (An epiphenomenon is a kind of side show to the main attraction. Steam is an epiphenomenon of boiling water; it has no existence in itself, and without the boiling water, there would be no steam. For many neuroscientists and philosophers of mind today, our consciousness is little more than a kind of steam given off by the brain.[16])

To dot the i's and cross the t's on this, let me say it in the simplest way possible. According to the most commonly accepted scientific view, *no one* wanted the Big Bang to happen. *No one* wanted the earth to form. *No one* wanted life to appear on the earth. And *no one* wanted life to evolve into us. There is no reason for any of it. It just happened.

This is a conclusion that the French Nobel Prize winning scientist Jacques Monod (1910–1976), one of the fathers of molecular biology, put perhaps more eloquently. Monod writes that:

> Chance alone is at the source of every innovation, of all creation in the biosphere. Pure chance, absolutely free but blind, at the very root of the stupendous edifice of evolution: this central concept of modern biology is no longer one among other conceivable hypotheses. It is today the *sole* conceivable hypothesis, the only one that squares with observed and tested fact. And nothing warrants the supposition – or the hope – that on this score our position is ever likely to be revised ... The universe was not pregnant with life nor the biosphere with man. Our number came up in the Monte Carlo game.[17]

From this insight, which is radically opposed to any kind of 'anthropic principle', Monod developed a rather bleak picture of our position in the world. 'Man', he tells us, 'at last knows he is alone in the unfeeling immensity of the universe, out of which he has emerged only by chance. His destiny is nowhere spelled out, nor is his duty. The kingdom above or the darkness below; it is for him to choose.'

Although his countryman Jean Paul Sartre (1905–1980), the most well known existentialist, had a low opinion of science – according to his lover Simone de Beauvoir, Sartre 'flatly refused to believe in science' and believed that 'microbes and other animalculae invisible to the naked eye didn't exist' – he nevertheless agreed with Monod on at least this proposition.[18] For Sartre, man has existence, but no essence ('his destiny is nowhere spelled out, nor is his duty'), and is ultimately a 'useless passion'.[19] Like Monod, Sartre believes that we must face this grim situation stoically and make the best of it, but there is absolutely nothing we can do about it. Suffice it to say, neither Monod's view nor Sartre's is one that sits well with our being caretakers or repairmen of the cosmos.

Matters dark and meaningless

Aptly, one character who shares Monod's gloomy vision of a chance-ridden universe and of ourselves as purposeless creatures within it, is the American horror fiction writer H.P. Lovecraft. Although Lovecraft was not a scientist (he was, though, a keen amateur astronomer) throughout his short life he professed an astringently materialist view of life and the cosmos. To a correspondent Lovecraft wrote:

> I am an absolute sceptic and materialist, and regard the universe as a wholly purposeless and essentially temporary incident in the ceaseless and boundless rearrangements of electrons, atoms, and molecules which constitute the blind but regular mechanical patterns of cosmic activity. Nothing really matters, and the only thing for a person to do is to take the artificial and traditional values he finds around him and pretend they are real; in order to retain that illusion of significance in life which gives to human events their apparent motivation and semblance of interest.[20]

Note that for Lovecraft, we maintain our 'illusion of significance' by maintaining values that are 'artificial' and which we only 'pretend' to be real. For anyone who embraces the belief that pure chance is responsible for our existence – and that includes quite a few of the most prestigious minds of our time – it logically follows that 'nothing really matters' as our actions can have absolutely no effect, one way or the other. It is difficult to see how values like love, freedom, truth, justice, beauty and others, that we hold give meaning to life, can have emerged from an existence accounted for by the 'blind but regular mechanical patterns' of 'electrons, atoms, and molecules', a vision of things that goes back to the pre-Socratic Greek philosophers Democritus and Leucippus. In such a world, values can have only a subjective and consensual existence, as fictions we agree on maintaining, as the only real things are purely physical, and, so far as we know, beauty, love, and other values are not made of atoms or molecules. If values are real, they exist in some non-physical kind of reality, of the kind Plato had in mind when he spoke of the good, the true, and the beautiful. Although the conclusion, that the values which give life meaning are really illusions, follows from the premise that chance is responsible for life and the universe, it rarely gets mentioned by the scientists who accept that premise.

Most of Lovecraft's weird fiction (of which I am a great fan) was published in the pulp horror magazines of the 1920s and 30s, of which *Weird Tales* is the best known. But while other *Weird Tales* writers, like Robert E. Howard (the creator of Conan) and Clark Ashton Smith, still have readers (I myself am occasionally one of them), Lovecraft's frankly overwritten stories acquired a serious critical cachet denied his friends and colleagues. Perhaps understandably, this critical importance was first recognised in the 1950s by the French, who a century earlier had embraced Edgar Allan Poe, a major influence on Lovecraft, when he was ignored by his fellow Americans.[21] Lovecraft's acceptance by the French, I would argue, had something to do with the bleak vision his stories portray, which is in essence the same as that of Sartre's grim philosophy and Monod's biological lottery. It is this atmosphere of existential dread, of some dark and terrifying *knowledge* breaking into our consciousness, that gives Lovecraft's tales a flavour as unmistakable as Kafka's, and which he shares with Sartre, whose existentialism Lovecraft would no doubt have turned his nose up at. For both the protagonists of Lovecraft's stories and Sartre's novels, knowledge is a trigger for a cosmic pessimism.

Nausea

For Sartre this cosmic pessimism is the 'nausea' of his most famous work, his first novel *Nausea* (1938), which I first read as a teenager. The hero of this novel has come to the startling recognition that things *exist*. But their existence has nothing to do with him, or with the stories or ideas he tells or has about them. They exist aggressively, in their own right; the names and categories and meanings we usually use to understand them – tree, stone, cloud, star – are simply falsehoods we tell in order to keep their strangeness at bay.[22] But now he *knows*, and the knowledge paralyses him. At one point in the novel, Roquentin, the protagonist, looks at the root of a chestnut tree and is perplexed by it. 'I no longer remembered that it was a root' he tells us.[23] Its existence frightens him. Like everyone else, he had taken existence for granted, and now it suddenly presses in upon him. It has lost its 'harmless appearance as an abstract category'. It had become the very 'paste' of things, and he cannot get away from it. At another point in the novel he is about to open a door when he looks at the strange thing in his hand and has no idea what it is. It was the door knob. Sartre's 'nausea' is not unlike states of mind associated with schizophrenia, when the connection between perception and feeling is unhinged. It is also suggestive that much of the inspiration for *Nausea* came from a bad mescaline trip Sartre had in 1936, in which he was attacked by devil fish and followed by an orang-utan, and in which umbrellas turned into vultures and shoes into skeletons.[24]

The misanthropic cosmological principle

Lovecraft's protagonists are also discomfited by knowledge. But while for Sartre a root or a doorknob spells doom, Lovecraft's dark insights are occasioned by more eccentric items. What knowledge means for Lovecraft can be best expressed by quoting the opening paragraph of his most famous story, 'The Call of Cthulhu', first published in *Weird Tales* in 1928. 'The most merciful thing in the world', Lovecraft writes:

> ... is the inability of the human mind to correlate all its contents. We live on a placid island of ignorance in the midst of black seas of infinity and it was not meant that we should voyage far. The sciences, each straining in its own direction, have hitherto harmed us little; but one day the piecing together of dissociated knowledge will open up such terrifying vistas of reality, and of our frightful position therein, that we shall go mad from the revelation or flee from the deadly light into the peace and safety of a new dark age.[25]

The theme linking the stories making up Lovecraft's 'Cthulhu Mythos', a literary mythology that Lovecraft's fellow *Weird Tales* writers contributed to and to which contemporary writers still add today, is that in dim ages past, well before man appeared, the earth was inhabited by strange, monstrous creatures, The Great Old Ones, who were expelled from it but who 'yet live on outside ever ready to take possession of this earth again.'[26] The Great Old Ones are terrifying indeed. Cthulhu him – or it – self is usually described as 'blasphemous', 'eldritch', 'loathsome', or another string of evocative adjectives, and is usually depicted as a kind of winged, tentacled, squid-like monstrosity of enormous size, who resides in the lost city of R'lyeh, sunken beneath the South Pacific. But while the actual beings of the Cthulhu Mythos – Yog-Sothoth, Nyarlathotep, Shub-Niggurath, 'the black goat of a thousand young', and the rest – are disturbing indeed, what is truly frightening about Lovecraft's cosmos is that these entities are *not*, as in traditional horror tales, supernatural, but merely products, like ourselves, of the chance work of accidental evolution in a 'wholly purposeless' universe. In our case, the 'boundless rearrangements of electrons, atoms, and molecules' that constitute the 'blind but regular mechanical patterns of cosmic activity' gave rise to us, and our 'artificial values' that allow us to live give us the false idea that, all in all, the cosmos is a relatively friendly and cosy place. But there we're wrong. The same blind forces that gave rise to us – and to Beethoven, Plato, Leonardo da Vinci, the Buddha, and Mother Teresa – also gave rise to these loathsome beings for whom we are negligible insects, when we are not mindless slaves or a tasty *hors d'oeuvre*. What is scary in the best of Lovecraft is this sense that we are ignorant of the truth about reality – like Sartre's Roquentin – and if we only *knew*, we would be afraid.

Lovecraft called his philosophy 'cosmicism', by which he basically meant that if we truly grasped the size, age, and sheer strangeness of the universe – an idea of which I tried to present earlier in this chapter – we would recognise that human life can play no important part in it, and that we are only temporary residents on a planet whose previous occupants are planning to return.[27] Possibly the earliest proponent of 'cosmicism', although he didn't use the term, was H.G. Wells (1866–1946), whose novel *The War of the Worlds* (1898) tells us that our world was 'being watched keenly and closely by intelligences greater than man's' – the Martians – 'and yet as mortal as his own ... minds that are to our minds as ours are to those of the beasts that perish, intellects vast, cool, and unsympathetic...'[28] When Lovecraft wrote 'The Call of Cthulhu', no one had yet thought of a Big Bang – the astronomer Fred Hoyle, an opponent of the idea, coined the phrase in 1949 – although Einstein's relativity was seeping into popular consciousness and quantum theory was raising its head. One can imagine contemporary Lovecrafts having a field day with our current cosmologies. And we can say that if there is an 'anthropic cosmological principle' that suggests human life – or at least life like ours – is somehow necessary to the cosmos, Lovecraft's 'misanthropic principle' suggests the exact opposite.

Of course not all fiction written from a 'cosmic' view is as dark as Lovecraft's. His contemporary Olaf Stapledon (1886–1950), for example, took a similar theme, yet didn't use it to scare his readers. Stapledon's *Last and First Men* (1930) and *Star Maker* (1937) are vast, cosmic future histories, in which different races, species, planets, and galaxies, arise from and sink into the cosmic depths. But reading Stapledon produces a sense of wonder and exhilaration, not terror. To be fair to Lovecraft, in his last work he too begins to see that the awareness that it's a big universe can lead to some insights more productive than the need to 'flee' into the safety of a 'dark age', as his story 'The Shadow Out of Time' (1936) makes clear. Some of his other misanthropic views also seemed to have softened with time. Sadly, by this time, Lovecraft was dying, and the insight came too late for him to make much use of it.

It's a tough cosmos out there

Mention of a 'misanthropic cosmological principle' leads us to an insight into Lovecraft's character. He was by all accounts an eccentric type, and for most of his life he lived in Providence, Rhode Island, where he was born, doted on by two aunts and surrounded by the remnants of his childhood. A brief marriage and life in New York proved disastrous. The asexual Lovecraft was not cut out for married life, and the immigrant population of New York offended his prim sensibilities. Lovecraft always fantasised himself as an aristocratic English gentleman of the eighteenth century, and to share the streets and subways of Brooklyn with Jews, blacks, Italians, Spaniards, and who knows who else was an affront to him. This aristocratic self-image was associated with a sense of an intellectual, or at least a cultural and aesthetic superiority. Lovecraft loathed the modern world, and as one critic has pointed out, we can read his ghoulish stories as a full out assault on it.[29] *He* knew that the values that make life meaningful for most of us were sheer illusions. *He* knew that our sense of being at home in the world was born of sheer ignorance. But unlike the ignorant fools who needed God or some other supernatural reality, and who believed that meaning and purpose were at work in the universe and not mere chance, he was tough enough to face this truth. He detested those fools who weren't, and so he wrote stories of cosmic terror to scare them.

The idea that those who embrace chance as the sole force at play in existence are tough enough to do without the illusions the rest of us enjoy, is a theme that comes up again and again. Even Lovecraft himself was accused of not quite making the grade. Of one story, 'The Whisperer in the Darkness' (1931), his biographer,

S.T. Joshi, remarks that in it, Lovecraft couldn't 'quite bring himself to admit that human penetration of the unknown gulfs of the cosmos is anything but an appalling aberration.'[30] Often those who reject meaning and purpose accuse those who look for it of wanting the world to be that way – who wouldn't, they concede – but of not being strong (or honest, or hard, or brave, etc.) enough to face the truth. But it strikes me that the opposite can be just as true, although it rarely gets a mention: that those who embrace meaninglessness and chance want to be seen as tougher (or more intelligent, honest, brave, etc.) then those who 'need' meaning and purpose. Frequently the search for or expectation of meaning is seen as a weakness. Yet again, the opposite can be just as true. The embrace of meaninglessness can be an expression of a hunger for superiority, the need to feel more intelligent and strong than the rest of us fools, just as it can be seen as a form of misanthropy, of a dislike of human beings. Both motives seem to me to be at play in the work of John Gray, considered one of the most important social philosophers of our time.

Shades of Gray

As in the case of Stephen Hawking's pronouncements on how the universe began, I find myself perplexed as to why John Gray's philosophy has acquired the aura of importance it has, at least for some readers. Prestigious names have sung his praises. For the late novelist J.G. Ballard, Gray's *Straw Dogs: Thoughts on Humans and Other Animals* (2002) was 'the most exciting book since Richard Dawkins' *The Selfish Gene*'. James Lovelock of Gaia fame tells us that Gray 'forces us to face the mirror and see ourselves as we are'. And for the critic Bryan Appleyard, whose opinion on other matters I've found cogent, Gray's book was 'unquestioningly one of the great works of our time'. Others have expressed similar appreciation. It may seem aberrant to fly in the face of such universal celebration, but as far as I can tell from reading Gray's books, he is basically a misanthropic pessimist, whose pro-nature and pro-animal remarks express little more than an emotional dislike for human beings. This misanthropy is something we've already seen in Lovecraft and is obvious to anyone who reads Sartre, his championing of human freedom notwithstanding.

I say I am perplexed by the importance Gray is given, because his philosophy seems as grim as Schopenhauer's, who also believed existence was meaningless, and expresses a ferocity toward human values reminiscent of the Marquis de Sade. I would even go as far as to say that remarks similar to Gray's anti-human rhetoric have been used in other contexts to justify murder, although I don't believe Gray himself intended his own comments for that purpose. Yet Gray's dislike of human beings leads me to suspect that he would not be too troubled if some of them quietly disappeared. Sadly, his animosity toward humans is predictably welcome because of our environmental concerns and guilty conscience about the planet. We have such a bad conscience about ourselves that one could say practically anything negative about human beings and be applauded for it as a deep thinker. Concern for the planet, however, can lead to some troubling places.

In *Straw Dogs* Gray remarks that from 'Gaia's', or the earth's, point of view 'human life has no more meaning than the life of a slime mould'.[31] When I wrote an article about Gray's idea some years ago, I pointed out that a similar assessment of human importance was championed by Charles Manson, currently serving a life sentence for the murders of Sharon Tate and Rosemary and Leno LaBianca in 1969.[32] Of the many pseudo-philosophical remarks Manson made and which were taken seriously by otherwise intelligent people, one was that a scorpion's life was more important than a human's.[33] While in prison, Manson had time to reflect on this insight, and to elaborate on its application. According to Manson, people and the 'system' were killing the planet. When Manson's 'Family' killed Sharon Tate – eight months pregnant – as well as Leno and Rosemary LaBianca, according to Charlie they 'gave their lives' and 'took lives' in order to 'clean up ATWA ... the whole life of the earth, in love and concern for brothers and sisters of soul'.[34]

ATWA is Manson's acronym for Air, Trees, Water, Animals (sometimes All The Way Alive), the name Manson gave to his radical 'ecological movement', shepherded by himself and two members of his Family, Lynette 'Squeaky' Fromme and Sandra

Good.³⁵ It may seem unbelievable, but as Nicholas Goodrick-Clarke points out, in recent decades Manson has been re-invented as an 'eco-warrior', and much of his rhetoric is eerily resonant with that of influential eco-groups such as Earth First!, the Animal Liberation Front, the Greens, and a good portion of New Age philosophy.³⁶ The Earth First! founder Dave Foreman declared that 'we are all animals' and agreed with Manson that human life is of no particular importance. 'An individual human life has no more intrinsic value than an individual Grizzly Bear (indeed, some of us would argue that an individual Grizzly Bear is more important ...).'³⁷ Although Gray would probably agree with this, I don't believe he would go to the lengths Manson did to make his point. But some of Gray's remarks make clear that being nature-oriented is not all sweetness and light. Whether we want to recognise it or not, there is a dark side to Gaia.

Gray's central idea is that 'humans think they are free conscious beings, when in truth they are deluded animals,' a borrowing from Friedrich Nietzsche (1844–1900), who, in his *Genealogy of Morals* (1887) said that 'man is the sick animal'.³⁸ (Nietzsche's Zarathustra called us the 'cruellest animal' as well, a sobriquet with which Gray would no doubt also concur.)³⁹ Free will, morality, and other specifically human concerns are for Gray simply illusions, just as they are for H.P. Lovecraft and the neuroscientists who believe consciousness is merely an epiphenomenon. We are, for Gray, fundamentally a rapacious species, intent on eradicating other forms of life, and we should own up to this, and forget all the nonsense about being anything other. Rather than call ourselves *homo sapiens* Gray suggests *homo rapiens*. Of course, not everyone agrees with this, and not all critics of this position are weak-kneed spiritual types. At the New Humanist website – where one finds 'ideas for Godless people' – Raymond Tallis, a secular humanist philosopher, takes Gray to task for arguing that we are nothing more than animals, a symptom of what Tallis calls 'Darwinosis'.⁴⁰ A similar sensibility informs the historian of science Kenan Malik's book *Man, Beast, and Zombie* (2001). Yet both Tallis and Malik would, I suspect, be surprised to find themselves in the company of a spiritual thinker like the poet Kathleen Raine (1908–2003), who agrees with them and against Gray that 'nowadays the term 'human' has been inverted to the point of signifying precisely what is least human in us, our bodily appetites...and all that belongs to natural man.'[41]

All our problems, and those of the planet, according to Gray, stem from our misconceptions about ourselves, and from our inveterate fantasy about being able to 'transcend' the 'human condition', which, as far as I can tell, is for Gray of the 'only human' character spelled out in the Introduction. Gray apparently has no capacity to grasp human greatness, and any reference to it is merely the cue for some remark to cut us down to size. What Maslow's 'fully human' would elicit from him I can only guess. 'A glance at any human,' he tells us, 'should be enough to dispel any notion that it is the work of an intelligent being'.[42] This remark would not be out of place in the writings of the Romanian arch-pessimist and one-time fascist enthusiast Emil Cioran (1911–1995), the titles of whose books – *The Trouble with Being Born*, *A Short History of Decay*, *On The Heights of Despair* – would not be incongruent with Gray's own.[43] 'In every man sleeps a prophet, and when he wakes there is a little more evil in the world,' 'By all evidence we are in the world to do nothing,' 'So long as man is protected by madness, he functions, and flourishes' – these are all taken at random from *A Short History of Decay* (1949). Cioran's misanthropic reflections are on first glance effective, because their aphoristic style has a shock effect, like a sharp jab to the ribs. Prolonged reading, however, reveals their basic hollowness. The same, I think, can be said of Gray's work. Cioran's misanthropy and cynicism about human values paved his way to a profoundly anti-democratic political philosophy that had no qualms about eradicating undesirable human beings, like Jews. The flipside of not liking human beings is not always saving the planet.

What seems to raise Gray's ire is the idea that we can in any way be 'masters of our destiny'. In 1957, Julian Huxley (1887–1975), brother of Aldous Huxley, and one of the most important biologists of the twentieth century, said that the universe was 'becoming conscious of itself ... in a few of us human beings', and that we had been appointed 'managing director of the biggest business of all, the business of evolution ...' He made these remarks in an essay entitled 'Transhumanism', in which he expressed his belief that 'the human species can, if it wishes, transcend itself.' For Huxley this meant 'man remaining man, but transcending himself, by realising new possibilities of and for his human nature', a sentiment with which I, and this book, are in accord – although Huxley's term 'transhumanism' has been adopted by proponents of a 'man-machine merger' with which I am not in sympathy.[44] Huxley even spoke of 'the cosmic office' to which we find ourselves 'appointed', a phrase with obvious similarity to the idea of ourselves being cosmic caretakers. Gray will have none of this and would, I suspect, greet any idea that we can direct our evolution with derisive scorn. Like Jacques Monod and Lovecraft, Gray sees little but chance at work in ourselves and the world. Not surprisingly, at the beginning of his chapter headed 'The Human' in *Straw Dogs*, Gray quotes Monod on mankind's desperate efforts to deny its 'contingency', a favourite word of Sartre's, which expresses, for Sartre at least, the fact that we are in no way essential to the world.

Gray's drift

According to Gray, 'epidemiology and microbiology are better guides to our future than any of our hopes and plans.' We are, he tells us, only a 'current in the drift of genes'.[45] 'Drift' is a favourite term of Gray's. He uses it frequently in a more recent book, *The Immortalization Commission* (2011), a bewildering attack on the idea that we can in any way 'cheat death', a theme already well explored in Ernst Becker's equally dour *Denial of Death* (1973). Gray repeatedly tells us about the need to learn the 'lesson of Darwinism', a lesson, apparently that Darwin himself didn't quite accept. 'Darwin never fully accepted the implications of his own theory' – an example of Gray being more royal than the king, of his 'cosmic toughness', and a reference to Darwin's concerns about the human consequences of his 'dangerous idea'.[46] Alfred Wallace, who arrived at the theory of evolution independently but at the same time as Darwin – and who was also one of the first proponents of an anthropic cosmological principle – was 'highly credulous', because he 'concluded that the human mind could not have developed simply as a result of evolution.' Wallace believed in a 'non-human intelligence' and a 'spiritual world' – he was interested in spiritualism and psychic phenomena – and so must of course not have had the nerve to accept the truth.[47] (Freud, for his part, could never 'cure himself of his fascination with telepathy'.[48] Once again, even a reductionist like Freud was not *quite* reductionist enough for Gray, and was in need of a 'cure' for his aberrant interests.) What is important for Gray isn't whether Darwin's view of evolution, or the canonising of it by today's neo-Darwinians like Richard Dawkins, is true, or can account for all the facts, but its 'lesson', which is something we *should* learn, as if we were naughty children. That lesson is simple. Genes drift, like sand dunes, and the difference between a slime mould, a scorpion, and myself is merely the way the sand has shifted. In my case it produced me; in H.P. Lovecraft's, squid-faced Cthulhu. As human beings are the great vandals of the planet, the idea that we could in any way do anything other than 'go with the flow' infuriates Gray. The fact that only a human being can think that his life is less important than a scorpion's or a slime mould's seems to have escaped Gray. No scorpion or slime mould can think this, and one doubts that they would if they could.

The central argument against Gray's anti-human animus, however, is this, if it is the case that we can never be masters of our fate – if we can only 'drift' – then why bother to write a book about it? What possible good could reading Gray's books do, if they can't change our destinies, which, according to him, are written in terms of epidemiology and microbiology? If, as he tells us, 'humans are animals ruled by blind selection' and can never hope 'to control the process of evolution', why bother to point this out?[49] (Or better still, if that is the case, then how can we possible save the planet or 'do as Gaia says'?) Or is it that the genetic drift has simply made Gray write his books, as it makes his readers read them? But if that is the case, then drift is drift, no matter what shape the sand dunes are. They are all sand. Or is it that in Gray's case, the drift has miraculously drifted into a dune that happens to be true?

This is the problem with all theories that want to explain human action and purposes in terms of blind, chance elements, whether it is 'genetic drift', 'the ceaseless and boundless rearrangements of electrons, atoms, and molecules', or 'the Monte Carlo game'.

If chance is behind everything, it is also behind the idea that it is behind everything, as well as the person or persons who believe that. Which suggests that, in all the possible arrangements of things, in *your* case, chance just happened to hit upon the arrangement that is true. But why should your chance arrangements of atoms or genes or anything else be anymore true than mine? The Nobel Prize winning molecular biologist Francis Crick (1916–2004) – co-discoverer of the structure of the DNA molecule – wrote that 'you, your joys and your sorrows, your memories and ambitions, your sense of personal identity and free will, are in fact no more than the behaviour of a vast assembly of nerve cells and their associated molecules'.[50] With all due respect, if that is the case, then it is also true of Crick's sorrows, memories, ambitions, identity – *and* his belief that these things are 'no more than the behaviour of a vast assembly of nerve cells and their associated molecules'. If my memories are 'no more' than molecules, and Crick's belief that this is true is also 'no more' than molecules, then why are his molecules 'more true' than mine? Molecules are molecules, just as sand is sand and drift is drift. The philosophers and scientists who make these kind of sweeping statements seem unaware that they undermine any possibility of their being true – which is the only reason we should give them a moment's thought.

More than animals

This sort of muddle is not the only thing wrong with Gray's argument. I say 'argument', but in fact, Gray does not argue. He simply asserts, and in our time, when it is ecologically-correct to have a guilty conscience about being human, making the kinds of statements he does has the *Zeitgeist* behind it. We feel so bad about being human that it is difficult to say anything positive about ourselves, or to question the kind of bad-mouthing we receive. I have found that if I do I am invariably accused of having something against animals and not wanting to save the planet, neither of which is true. In our anarchic times, filled with very real ecological, social, and political threats, it is satisfying to have a scapegoat on whom to pin the blame. For Gray that's us. But it is also dangerous;

we've seen how close some of Gray's remarks are to those of an ecologically-correct convicted killer. I agree that we face enormous challenges – I'd be an idiot if I didn't – and that many of our crises are of our own making and that we need to grow beyond our short-sighted selfishness. But I don't think throwing our humanity out with the bathwater is the best way to go about this.

One of Gray's mistakes is to believe that we were all happy with being animals until Christianity arrived. That is not true. As soon as we became self-conscious, we realised we were somehow *different* from the other creatures around us, and the history of the human mind as been, more or less, an unpacking of that intuition. For thousands of years, human beings have felt and acted as if they were something *more* than animals – every religion is, in some way, about this, and there were religions before Christianity. We always knew we had an animal nature – that isn't in doubt. But we didn't believe it summed us up completely. It is in fact only since the rise of science, and specifically of Darwinian evolution, and the crude, materialist reductionism born of them, that we have had the idea that we are only animals force fed us, with the best intentions, no doubt. Whether it is selfish genes or naked apes, it is only in the last few centuries that we have come to see ourselves as sick, deluded, cruel, rapacious – you can pick the adjective of your choice – animals, formed, not in the image of God, Nous, or the *Ein-Sof*, but of drifting matter. And as Kathleen Raine has pointed out:

> The world has never been more hideous, more
> uninhabitable, than the world created by an ideology
> that proclaims that this world is all, which gives to
> matter a primacy, an all-importance unknown to other
> civilisations'.[51]

With the corollary that, as Colin Wilson writes, speaking of today: 'it is probably true to say that there has never been a point in human history when humankind had a more depressing view of itself.'[52]

2. A Dweller on Two Worlds

In the last chapter I tried to give an idea of the size, age, and sheer strangeness of the universe we live in, and to present some of the scientific accounts of how we and it got here, and the impact those accounts have had on our idea of ourselves and our place in the cosmos. But before I continue I should clarify one point. Although we use the terms interchangeably, there is a difference between a 'universe' and a 'cosmos'. We use 'universe' or 'world' to refer to everything, to the 'all', or to the 'space' in which the galaxies, stars, planets and other stellar objects exist. Our word 'universe' comes from the Latin *universum*, which can be understood to mean 'the one turning' or 'the one rotating'. Cosmos, however, has a more specific meaning. It comes from the Greek *kosmos*, which means an orderly or harmonious system, and is the opposite of *khaos* – our 'chaos' – which for the Greeks meant a formless void, the state of existence before creation. This is the *tohu bohu* or 'trackless waste and emptiness' of Genesis, before God said 'Let there be light'. It is also the 'grim darkness' of the Hermetic creation myth, and the 'void' created by the *Ein-Sof*'s 'contraction' in the *tzimtuzm*. (We can also think of the 'pre-existing vacuum' in which, according to Stephen Hawking, a 'quantum fluctuation' led to the Big Bang .) The idea of harmony and order, or beauty, in relation to a cosmos, comes through in our word 'cosmetic', which in the original Greek meant something that beautifies, or enhances beauty, such as an ornament, but which in our time has come to mean a superficial or false kind of beauty. We think of cosmetics as 'make up', something which provides glamour, but only deals with surfaces, not reality, and which is then considered false or not essential. (The fact that when we tell a falsehood, it is something we 'made up', just as when we pretend or invent something we 'make it up', also seems suggestive.)

The first person to speak of the world as a cosmos was the Greek mystic-philosopher Pythagoras, who lived in the sixth century BC. Most of us remember learning the Pythagorean Theorem in school, in which, in a right-angled triangle, the area of the square of the hypotenuse, equals the sum of the area of the squares of the other two sides ($a^2 + b^2 = c^2$), and many of us cannot forgive Pythagoras for discovering it. In the last chapter I mentioned that Pythagoras believed that the world was made of numbers. What this means is that Pythagoras recognised that there was an order within nature and the world, and that this order is mathematical, an idea that can be found championed today in the work of the scientist Roger Penrose.[1] Pythagoras is also credited with discovering the musical octave, the fact that different musical notes are related to each other harmoniously because there is a simple ratio between the notes and the length of string making them. It was from this insight that Pythagoras arrived at the notion of the 'music of the spheres', the idea that the planets and stars moved according to a similar mathematical equation, and in doing so produced a kind of cosmic harmony. It was this recognition of an order and of the beauty that it produced, that led Pythagoras to call the world a cosmos.

Although scientists still use the word, and speak of 'cosmology', the kind of order and beauty that Pythagoras recognised, and which informed our ideas about the world for most of human history – a similar notion can be found in cultures other than in the west – is no longer the way in which we perceive the universe. With Big Bangs, black holes, colliding galaxies, and dark matter, if there is a music of the spheres today, it must be much more along the lines of atonal music – Arnold Schoenberg comes to mind – or Stravinsky's raucous *Rite of Spring*, than Bach. Perhaps the aleatoric or 'chance' music of John Cage is even more apt. Although we may find a strange beauty in photographs of distant galaxies and exploding stars, the universe we inhabit is much more turbulent, explosive, and eruptive, than the one Pythagoras knew. And as our own earth-bound world is also more turbulent than the one Pythagoras knew – the sheer increase in population and the complexity of life suggests this – we can perhaps find new meaning in the old Hermetic axiom 'as above, so below'.

The end of beauty

It is curious that in many ways, changes in our ideas about art and beauty have paralleled changes in our ideas about the universe. In postmodern and conceptual art, beauty itself is considered a false idea, merely 'cosmetic'. The idea of a work of art having an immediate sensuous impact is suspect. I should say an art 'piece', as it is usually referred to today, emphasising its broken or fragmented character.[2] Today art, like life, consciousness, and the cosmos, must be 'explained', hence the often obscure writings that accompany much contemporary art, and which are also equally in need of explanation. As the psychiatrist and philosopher Iain McGilchrist remarks, 'it was not the norm, until the advent of modernism, for people to find new styles of music unpleasant or incomprehensible', thus exploding the myth that if we find some piece of postmodern music unlistenable, then it must be a work of genius, because all new works must sound strange to mere philistines.[3] Nearly half a century ago, Kathleen Raine lamented that in the modern world, 'the word beauty and the idea of the beautiful, has ceased to count for anything,' a sentiment echoed by the cultural historian Jacques Barzun, when he wrote that 'not until the flowering of modern art did uncontrolled rage in vituperation and...spluttering sophomoric insult ... find their way into works of artistic pretensions ...'.[4] Barzun's remark was in response to the work of Alfred Jarry, creator of the nihilistic *Ubu* plays, paragons of what the surrealist André Breton called 'black humour', and his slogan that 'We will not have brought down everything in ruins until we also destroy the ruins'.[5] A look at the unmade beds, dissected sharks, and canvases smeared with animal dung that adorn our museums and galleries suggests that the profitable work of bringing down the ruins is still going on. Echoing Raine and Barzun, McGilchrist concurs that today 'beauty has been effectively airbrushed out of the story of art' and that it has become 'unsophisticated' to speak of it.[6]

At the risk of being unsophisticated, I will say that for there to be a cosmos, there must also be beauty, an insight recognised by the Russian religious existentialist Nikolai Berdyaev. In *The Meaning of the Creative Act* (1916), an exhilarating book sadly little read today, Berdyaev wrote that 'in every artistic activity a new world is created, a cosmos, a world enlightened and free.' Even more, Berdyaev argued that 'the cosmos is just this: beauty as being'. 'Cosmic beauty,' he tells us, 'is the aim of the world process: it is another kind of being, a higher being which is in the process of creation'.[7] Neither Steven Weinberg's purposeless universe nor the one that emerged from Stephen Hawking's vacuum can be said to be a cosmos, at least not in the sense that Berdyaev means. And although Berdyaev and Pythagoras are in many ways worlds apart – the one an ex-Marxist Christian existentialist, the other a pre-Socratic mathematical mystic– they both share the sense that, for there to be beauty in the cosmos – which means that for there to be a cosmos at all – man must be present. For Berdyaev, man is a creative agent at work in the world, and his work is likened unto God's. For him, 'the cosmos is being created; it is not given, it is a task set', and it is we who must complete it.[8]

Chaoskampf

From the little we know about him, Pythagoras seems a less romantic, less dramatic character than Berdyaev. He was much more of a scientist, and he shares the ancient belief in a cosmos already completed and perfect, unlike Berdyaev, whose philosophy of human freedom and creativity is much more aligned with the Kabbalistic idea of *tikkun*. Yet this perfect, harmonious cosmos would be unknown, were it not for the mind that knows it, a mind that, in order to know it, must in some way be of the same character as the cosmos. It too must have the same order and beauty that Pythagoras recognised in the cosmos. We can say then, that for Pythagoras, if there is no mind, there is no cosmos. This equation can be seen in the many creation myths in which a hero-god battles a 'chaos monster', a struggle that in comparative mythology has come to be called by the German term *Chaoskampf* or 'chaos struggle'. Elsewhere I have written about this in the context of Egyptian mythology and the sun god Ra's battle with the demon of chaos Apophis.[9] The struggle depicts both the emergence of order out of chaos, and the rise of a self-conscious, individual 'I' out of what the historian of the psyche Erich Neumann called the 'ouroboric' condition, named after the Ouroboros, an alchemical image of a snake biting its own tail, a symbol of a self-enclosed unity. This 'chaos struggle', this rising of order out of a undifferentiated state, is something we all go through in our third year of life, when we break out of the ouroboric egg of infancy into the light of being an individual 'I'. This is why we speak of the 'terrible twos', because it is at that age that we begin to gain a separate identity, which is expressed in our ability to say 'no', a difficult development, as every parent knows.[10] It is at this point too, that a 'world' emerges for the first time, for it is then that we start to grasp the difference between our own self and the surrounding environment, in which, up until then, we had been completely immersed. Up until then, for us, there was no 'world' because there was no 'self' to be set apart from it. So, at a fundamental level, the emergence of a world, a cosmos, out of chaos, depends on the emergence of an 'I', a mind, out of an unconscious, undifferentiated state.

Lost in the cosmos

Are we justified then in asking if the purposeless universe emerging from Hawking's pre-existing vacuum can be considered a cosmos? The question is not whether the Big Bang and other mysterious features of this scenario are true, but whether such a scenario can give rise to the order and harmony that Pythagoras and other ancient cosmologists recognised as essential to a cosmos. We know that the pre-Copernican world, which lasted until only a few centuries ago – until 1543 in fact, when Copernicus published his *On The Revolutions of the Heavenly Spheres* – was indeed a cosmos. Copernicus changed everything when he showed the earth and the other planets revolved around the sun, and not, as had been believed, that the sun, planets, and everything else revolved around us. For pre-Copernican man, the fixed stars suggested an eternal order, and the celestial crystalline spheres, each carrying a planet in its graceful orbit around the earth – the centre of everything – made, as Pythagoras believed, a beautiful cosmic music. In that world, the eternal order of the heavens was reflected in the life of man on earth, and in this way, the lower, sub-lunar material plane, participated in the higher regions, which were believed to have been made of a more spiritual substance. Today we know that the matter here on earth is the same as that of the stars, and that the stars are subject to the same physical laws as the earth is. From this perspective we can turn the Hermetic axiom around this time and say 'as below, so above'.

An early sign of the disquiet that comes to the unfortunate protagonists of H.P. Lovecraft's fiction when they realise just how big the universe really is, can be found in an oft quoted remark by Pascal, who we mentioned briefly in the previous chapter. Pascal was a young contemporary of René Descartes, and was there, we can say, at the beginning of our modern world.[11] Regarding the vast, strange universe that Copernicus' theory revealed to us, in his posthumously published *Pensées* (1669) Pascal wrote that 'the eternal silence of these infinite spaces frightens me'. Note that Pascal's emphasis is on *silence*. As I've argued elsewhere, it was not merely the size of the new Copernican universe that troubled Pascal. The Alexandrian Hermeticists spoke of a 'space beyond heaven where there are no stars', a perhaps early intuition of the notion of an 'observable universe', and the Renaissance Hermeticist Giordano Bruno argued for an infinite universe, filled with innumerable worlds.[12] Both ideas suggest a large universe, one much bigger than the reigning Ptolemaic one. (Ptolemy – AD90–168 – was an Alexandrian astronomer whose complicated system of 'nested' planetary spheres dominated astronomy until the arrival of Copernicus.) What frightened Pascal is that this new universe seemed, as far as he could tell, oblivious of him. It was alien, and its reality seemed to negate him, much as the reality of a door knob or tree root negates the hero of Sartre's *Nausea*. These vast, empty depths seemed to have nothing to do with him, and they left him floating, lost in space. More than two centuries later, in 1885 Nietzsche caught the tone of Pascal's cosmic fear when he wrote in his notebooks that 'since Copernicus man has been rolling from the centre toward X.'[13]

Everybody knows this is nowhere

A more contemporary soul who shared Pascal's cosmic disquiet is the mystical poet and diplomat O.V. de Lubicz Milosz (1877–1939). Milosz is a remarkable character. Of Polish, Lithuanian, Jewish, and Italian descent, Milosz wrote in exquisite French – he lived much of his life in France and is considered a French poet – and his polyglot talents were put to good use in the years he spent as a member of the Lithuanian delegation to the Peace Conference of 1919, following World War I. Later he was the Lithuanian representative to the French government and then the Lithuanian delegate to the League of Nations. In 1931, in recognition of his service, he was made a chevalier of the Légion d'Honneur. For most of his early poetic career, Milosz was a typical decadent of the *fin-de-siècle*. At one point he was an acquaintance of Oscar Wilde during the disgraced playwright's last days in Paris. The ennui and febrile atmosphere of the decadents, however, exhausted Milosz, and in 1901 he fumbled a suicide attempt. The major crisis of his life came in 1914, when a profound mystical experience led to an audience with 'the spiritual sun', an encounter he shared with his fellow poet, William Blake, of whose work Milosz was unaware. Blake was a reader of Swedenborg, and Swedenborg it seems had a similar experience himself. Blake told his friend Crabb Robinson that he had 'conversed with the spiritual sun' on Primrose Hill in North London.[14] And in *The True Christian Religion* (1771), Swedenborg tells us that 'The Lord is the Sun of the angelic heaven, and this sun appears before the eyes of angels when they are in spiritual meditation. The same thing happens with a man in this world, with whom the Church abides, as to the sight of his spirit'[15].

In his encounter with the 'spiritual sun' Milosz found himself 'soaring freely through space' until he landed on a mountain peak, which itself continued to rise to the heavens. He reached a dense cloud, whose consistency he compared to 'the freshly discharged seed of man'. Then a gigantic reddish 'egg' was hurled towards him. It then became a golden lamp which grew in size until it had the shape of an 'angelic sun'. Hovering above his forehead, this 'sun' then looked deeply into his eyes.[16] A decade later Milosz published his first metaphysical work, a long Hermetic prose poem called *Ars Magna* (1924). *The Arcana*, another long mystical prose poem, followed in 1927. After this, Milosz abandoned poetry; he returned to it only once more, nearly a decade later, to write his last poem 'Psalm of the Morning Star' (1936).

Milosz was a lifelong reader of Goethe, Plato, and Dante, and his metaphysical poetry shows their influence. Although the exact metaphysics of *Ars Magna* and *The Arcana* is difficult, it is clear that Milosz has inherited the Romantic rejection of the materialist account of the universe. As he was for Blake and Goethe – who argued against his theory of optics – Isaac Newton (1642–1727) is Milosz's target. Or to be more precise, Newtonian space, the kind that frightened Pascal. It was with Newton that the idea of an 'absolute' space and an 'absolute' time came to dominate western consciousness. In this sense space becomes sheer empty distance, and time mere forward flow. Rather than a vast cathedral of celestial spheres, arcing over the earth, space since Newton has meant an abstract *extension* in all directions. Likewise time for Newton means eternal *succession*, a temporal parallel to infinity, 'one thing after another'. Both embody what Hegel called a 'bad infinity', a kind of straight line forever stretching out into emptiness (a 'good infinity' for Hegel was like a circle, forever returning to itself, like the alchemical Ouroboros). If we take the co-ordinates 'here' and 'now' and draw a line from them extending infinitely in any direction – which would take an eternity to do – we will never reach an end. If we continued that line for a billion years, the distance ahead of us would still be as infinite as when we began, just as the time remaining to us would still be eternal, with as much eternity left as when we started. If you think about this for a minute, Pascal's angst becomes quite understandable.

Milosz had his own frightful experience of Pascal's infinite spaces, although in his account, they were not silent, but filled with a barbarous cacophony. In a vision Milosz found himself on a vast, deserted plain. Overhead was a 'universe of terror', jammed with billions and billions of 'tormented cosmic systems' turning in endless rotation. Rather than a harmony of the spheres, these stellar wheels emitted an 'odious criminal noise, the enemy of all meditation, of all composure', rather like today's i-Pods. What these oppressive stellar perpetual motion machines said to Milosz was that 'you must multiply and divide the infinite by the infinite during an eternity of eternities'.[17]

This nightmare of an immense, threatening blackness in which vast cosmic systems turn incessantly reminds us of William Blake's dark visions of 'starry wheels' and, as mentioned, of Pascal's angst about infinite space. Repelled by this crushing cosmic futility, Milosz attempted to find the 'place' of mankind. In his dark vision Milosz had seen that our 'place' had been lost, annulled in the infinite emptiness of abstract space, where, in the words of one of his commentators, we 'proclaim ourselves sovereign for a day of a lump of matter sentenced to slow decay in the darkness of a death without beginning or end.'[18] We were, it seemed to Milosz, literally nowhere, or, if it was a 'where', it was hell.

The problem, Milosz saw, as did his predecessors Blake and Swedenborg, was that in the Newtonian world, we identify infinity and eternity with abstract extension and succession. But the truly infinite and eternal are outside space and time. By identifying the infinite with the dark emptiness of Pascalian space, we had literally lost our 'place' in the cosmos, because in a universe without borders, there can be no 'place', no difference between 'here' or 'there'. In Newtonian space, there is no 'up' or 'down', 'left', or 'right', 'forward' or 'backward'. And if we had lost our place, so had the cosmos. The question 'Where is space?' came to obsess Milosz, and he found he could not answer it. Given a Newtonian universe of abstract infinite extension, no answer seemed possible. In such a universe, there is no 'where' for space to be. Anything that could 'contain' space would itself only be more space. Everything is 'in' space, but space itself is 'in' nothing, it seemed. So wherever we were, it was, in a sense, no 'where'.

Universal Man

Milosz found help in asking his questions through a reading of Hermetic, Kabbalistic, alchemical, and mystical writings, which, he tells us, he was ignorant of until his experience. He also found some support in the work of Albert Einstein (1879–1955), which, in 1914, was still relatively new and obscure. Milosz tells us that he was as ignorant of Einstein's ideas then as he was of mystical literature. In Einstein's relativity, Milosz felt he had found a beginning of an answer, because Einstein made space and time *relative* to an observer- that is, to man. For Milosz this meant that although Copernicus had unchained the sun from the earth, and Newton had sent us drifting through absolute space, we were with Einstein back at the centre of things – at least the individual observer was.

But Einstein's ideas were only a mathematical, scientific rendering of truths central to the whole western esoteric tradition. As mere specks of matter, it is true that we are insignificant motes of cosmic dust, floating in empty space. But this is to see ourselves in our fallen nature. Blake and Swedenborg knew that the truly infinite and eternal are not 'there', *outside*, but 'here', *within* man, within his imagination, which for both, and for all poets and seers, is not the 'fancy' we refer to when we say 'that's just your imagination' – that is, something untrue or, as mentioned earlier, 'made up' – but the creative energy and origin of the cosmos itself. It is the imagination in the sense that Henry Corbin (1903–1978) writes of it when he speaks of an 'Imaginal World', the *mundus imaginalis*, a world of archetypes, forms, images, and intelligences which are neither purely physical nor purely ideal, but exist in an intermediate realm of their own. For Swedenborg, space and time are 'states', forms of consciousness, not absolute realities in themselves. When the doors of perception are cleansed, Blake famously said, we will see all things as they really are, infinite. 'Every space larger than a red Globule of man's blood is Visionary, and is created by the Hammer of Los: And every space smaller than a Globule of Man's blood opens Into an Eternity of which this vegetable Earth is But a Shadow' (*Milton*, 1810). We, then, are not 'in' space; space, and everything it contains, is within us, within our imagination.

It is in this sense that man is a microcosm, a little universe. Earlier I mentioned Swedenborg's idea of the Grand Man, a theme Blake continued in his 'human form divine'. It is the same as the Kabbalistic Adam Kadmon, who we met in the last chapter, and from whom the energies of the divine create the cosmos. It is also the same as the Hermetic belief that ultimately, the cosmos itself is in the form of man. Not man the animal, man the rapacious despoiler of nature, but the man of imagination, within whose infinite inner spaces the entire outer world exists.

What has happened, Milosz tells us, is that we have fallen from our original state. Like Hermetic man embracing nature, or the shattering of the *sephiroth*, for Milosz some primal cosmic crisis has brought us to this hell of empty space and time, this desert of mechanical, ever-turning starry systems, Steven Weinberg's purposeless universe. Somehow, through what Milosz calls 'Adam's prevarication' – that first sin in the Garden of Eden – we have been separated from the original spiritual universe, and we find ourselves here, stranded in a world of pointless matter. We have, in a sense, been turned 'inside out', and where, in our primal state, we were free, infinite beings, sharing in divine love and wisdom, containing a world, we are now mere lumps of flesh, lost in a dark, oblivious universe whose oppressive reality negates us. But as with Hermetic man and the Kabbalistic 'sparks', something remains of our original state, and from time to time a vague memory of where we have come from rises up in us, and we suffer from that ailment known to all true poets and philosophers, homesickness.

Adam Strange

When I was young one of my favourite comic book heroes was a character named Adam Strange who, aptly enough, appeared in a magazine called *Mystery in Space*. He was not a super-hero per se. He didn't have super powers, although he did sport a rocket pack and enviable ray gun, but his life was in some ways more fascinating than Superman's, Batman's, and the other heroes whose adventures I greedily absorbed. Adam Strange was an archaeologist, and on a trip to the Andes, while escaping from some Indians – admittedly he had discovered their secret treasure – a remarkable thing happened. Leaping across a chasm, Adam was hit by a strange ray – he later discovers it is called the 'zeta beam' – and finds himself transported to the planet Rann, an earth-sized world orbiting Alpha Centauri, the closest star to our own sun. In an instant, he is shot 4.3 light years – or 25 trillion miles – across space to another world. Adam meets a woman, has adventures, saves the planet, and becomes a hero. But then suddenly, the effect of the zeta-beam wears off, and he finds himself back on earth. But he discovers where and when the zeta beam will strike the earth again – it happens to be on the Malayan coast, south of Singapore – and he is determined to be there, and to return to Rann. And so it went, for the rest of the series: Adam Strange cutting through jungles or climbing mountains in order to catch the next zeta beam, only to be sent back to earth once again when its effects wear off. But he is determined to become a citizen of his new world, and one day, he will discover a way to remain on Rann forever ...

I don't know if Julius Schwartz, Gardner Fox, and Mike Sekowsky, who started the series, realised it, but they had hit on a perfect metaphor for man's dual nature: that he is a dweller on two worlds: one of everyday, commonplace fact, where he spends most of his time, and the other of something much more mysterious and adventurous, which he enters only briefly. It is of course a common theme in Romanticism but the fact that this insight had filtered down from 'high culture' to the lowlands of comic books, suggests that it is something inherent in us, perhaps even a Jungian archetype. (It also suggests once again that popular culture is often a medium for deep philosophical thought.)[19]

Planets of a double Sun

As I say, not surprisingly, the two worlds theme is something we find in Romanticism. Romantics, like no one else, are dissatisfied with the grim, 'real' world of mundane necessity and have a hankering for another world, a land more in harmony with, in Yeats' phrase, 'the heart's desire'. One of the most effective and entertaining expressions of the two worlds can be found in the German writer E.T.A. Hoffmann's glittering fable *The Golden Flower Pot* (1814). The theme runs through all of Hoffmann's work – as it did through his life – but in this tale, generally regarded as his greatest work, he embodied perfectly the enduring tension between the everyday world and that of 'magic'.

Hoffmann (1776–1822) was a curious soul, who seemed to live two radically different lives: by day a respected juror and Prussian civil servant, by night a poet, musician, artist, and critic. His stories, which read like a meeting between Hans Christian Andersen and Edgar Allan Poe, have never really made the transition to English, in spite of excellent translations, and for English readers he is most known today as the inspiration for Offenbach's *Tales of Hoffmann* and Tchaikovsky's ballet *The Nutcracker*.[20] Hoffmann would have been pleased with this; although most of his music is lost, Hoffmann composed some ten operas, as well as two symphonies, several cantatas and many other works. Music for Hoffmann was the clearest evidence that we are natives of another land. Hoffmann was one of Beethoven's earliest champions and was also central in promoting instrumental or 'absolute' music, at a time when vocal music was supreme. In an essay on Beethoven Hoffmann wrote: 'Music reveals to man an unknown realm, a world quite separate from the outer sensual world surrounding him, a world in which he leaves behind all precise feelings in order to embrace an inexpressible longing'.

We will return to this idea of an 'inexpressible longing', but it seems more than coincidental that Pythagoras should speak of a 'harmony of the spheres' and Hoffmann and innumerable others should regard music as a portal to 'other worlds'. In the kind of world fashioned by genetic drift and evolutionary necessity, music plays no such role, as evidenced by the Darwinian evolutionary psychologist Steven Pinker's remark that music is an indulgent but inessential by-product of adaptive behaviour, on a par with, and as meaningful (or less) as pornography and fatty foods.[21] (In Pinker's cosmos, then, we could say that we would hear the 'muzak of the spheres'.) Beethoven once asked their mutual friend Bettina von Arnim to tell Goethe 'to hear my symphonies and he will say that I am right in saying that music is the one incorporeal entrance into the higher worlds of knowledge'.[22] How Beethoven would have responded to Pinker's remark requires little imagination.

The plot of *The Golden Flower Pot* involves a young, poetic, but clumsy student, who is troubled by a recurring vision of a glittering green snake. The snake turns out to be Serpentina, the daughter of the Archivist Lindhorst, who is himself secretly an elemental salamander, exiled to earth because of a transgression committed ages ago, Hoffmann's allusion to the primeval 'cosmic crime'.

Anselmus, the student, is betrothed to Veronica, who is beautiful but worldly, and who wants him to become a Hofrath at court and lead a dutiful, sensible life. But the Archivist hires Anselmus to copy out a magical manuscript, which tells the story of Lindhorst's own fall from grace, and of the ongoing battle he continues against the forces of darkness. The tension of the narrative, which is full of initiatory trials and strange, mystical adventures, lies in Anselmus having to choose between the two worlds: that of the practical, down-to-earth Veronica, with whom he is bound to have worldly success, or the mysterious but tempting Serpentina, who makes him dream of strange but beautiful worlds. I can give only the barest outline here and readers are advised to enjoy Hoffmann's delightful and enlightening tale themselves. One of the important elements in *The Golden Flower Pot* is that, unlike earlier *Märchen*, or 'fairy tales' written by his fellow Romantics Goethe and Novalis, Hoffmann's takes place in an everyday world, Dresden, in fact. It is precisely the contrast between the world of magic (Serpentina), and that of the commonplace (Veronica), that gives the story its power. In the earlier tales, which take place entirely within a 'fairy land', the essential *opposition* between the two worlds is lost.[23]

A more recent but equally effective expression of the 'two worlds' principle can be found in the writing of the Scottish novelist David Lindsay (1876–1945), best known for his remarkable Gnostic work *A Voyage to Arcturus* (1920). In *A Voyage to Arcturus*, the protagonist, Maskull, is, like Adam Strange, transported from earth to another planet, this time Tormance, which orbits Arcturus, the brightest star in the constellation. Appropriately, Arcturus is a double star, and in choosing it as the setting for his masterpiece, Lindsay seems to have displayed an acute intuitive grasp of the 'two worlds' problem. As Colin Wilson remarks, all artists are planets of a double star: one is 'the world' which we accept as given and resign ourselves to, and the other is that strangely familiar but mysterious place for which we sometimes feel a poignant homesickness.[24] On Tormance Lindsay's hero Maskull sets out on a kind of Gnostic 'pilgrim's progress', encountering a series of strange beings and creatures all of whom claim to posses the answer to life – love, sex, power, knowledge – but who all in the end prove false. The 'muspel fire' that Maskull seeks, driven by the drum beats of Surtur – a symbol of the evolutionary urge to transcend himself – is a symbol of an almost total rejection of the world and its hollowness, an austere Buddhist denial of what those less fastidious than Maskull accept as reality, but which is itself a reality of uncompromising value and meaning. Again, the plot is simple, but a reader soon finds himself absorbed in Lindsay's utterly convincing imagination, his ability to create ever more fantastic encounters which at the same time embody profound philosophical questions.

It is, in fact, Lindsay's deep feeling for 'other worlds' and their metaphysical meaning, and his lack of sympathy for *this* one that is one of his problems as a writer. After *Arcturus*, which garnered some critical acclaim but little commercial success, Lindsay felt he needed to keep his novels more 'down to earth', in order to attract more readers, and so his remaining five metaphysical fantasies – he wrote one romantic pot boiler in a failed attempt to be popular – are all set on earth. Alas, while Lindsay can write powerfully about deep metaphysical dimensions, his facility to depict the everyday is severely limited, and these novels, too, were commercial failures. In his novels, *The Violet Apple*, *Devil's Tor*, *The Sphinx*, and the unfinished *The Witch*, there are long stretches of everyday reality which, to most readers, are barely readable; unlike Hoffmann, Lindsay seemed unable to bring them to life. But these stumbling excursions into the common light, which Lindsay mistakenly felt were obligatory, are more than made up for by the sudden eruption of the 'other world'.

Stairways to heaven

Perhaps this contrast between the everyday and the 'other world' is best depicted in Lindsay's second novel *The Haunted Woman* (1922). The heroine, Isbel Loment, is a woman of artistic temperament who leads an empty existence, moving from hotel to hotel as a companion to her aunt. She is engaged to a decent but ordinary man she does not really love yet can find no inner strength to break off the engagement and give her life some real direction. By chance she is invited to a strange house and there meets Henry Judge. Both discover that they have had a similar strange experience in the house. A mysterious stairway, apparently unknown to the other guests, has appeared to them, and when they ascend it they find themselves in a strange room, which has the inexplicable effect of transforming them into who they *really are*; it is a kind of 'self-actualising chamber', to see it in Maslovian terms. The conventions of everyday life, and the fears and feebleness that constrain us, have disappeared, and for a brief time they are both their 'true selves'. They agree to try to ascend the secret staircase again and meet in the room. They do and find that they are in love with each other, although our conventional idea of love does not convey the experience of spiritual attraction and the power of true 'meeting' that they undergo. In *The Witch* Lindsay wrote 'While bodies, in a physical world, might join, spirits never could. For bodies were in their own home, but spirits were exiles and captives in a strange land'. The passion that Isbel and Judge feel is not physical or even erotic, in a Platonic sense, but a hunger for *something else*, some reality their normal lives cannot provide. Out of the window of the room, in the distance, they hear a painfully beautiful music, and see a bright, spring sunshine, although in reality it is a misty autumn day. The music draws them, but somehow they cannot get to it, and when they leave the room and descend the stairway – which they inevitably must – they immediately forget their experience, and return again to their smaller, ordinary lives.

The stairway is a clear symbol of an ascent to a higher realm, and in the beginning of the book, there is a reference to music with which Hoffmann would have agreed. Isbel hears someone playing the opening of Beethoven's Seventh Symphony on the piano. The music, she feels, was like a curtain being drawn, 'revealing a new and marvellous world' and with the 'famous passage of gigantic ascending scales ... she immediately had a vision of huge stairs going up.'[25] Lindsay refers to the piece as 'that fragment of giant-land', and throughout his work, music plays an important role. A central motif in his last, unfinished novel, *The Witch*, is something he calls 'the third music', a metaphysical insight that cannot be expressed in words, but which is conveyed unmistakably in music.

The two worlds

I should at this point clarify what I mean when speaking of 'two worlds'. There is, of course, only one world. I don't mean planet or galaxy, but the reality in which we find ourselves. I cannot walk into another world in the same way that I can walk into another room or another house. I am here, in this universe, and there doesn't seem to be any exit doors, unless you consider death one, and we are not even sure about that. We are here, now, and no amount of wishing or willing will send us to Rann or Tormance or Mars, where John Carter goes in the Edgar Rice Burroughs novels I devoured in my childhood. The idea of an actual 'other' world we can escape to, is a literal interpretation of what is really a metaphysical distinction. Many a Romantic has taken to the road, in search of some other place, where life will be much better than it is wherever he started out from. The grass is always greener on the other side, until you get there and you see it is exactly the same shade as in your own garden. We all know that no matter where we go physically, 'we' are still there. If there is another world then, it isn't 'out there', but 'in here', and it consists of seeing this world in a different way than we 'normally' do. I put 'normally' in quotation marks because we will look at exactly how normal this is.

As Kathleen Raine put it: 'the imagination does not see different things, it sees things differently.'[26] This is why the meaning content of many mystical experiences is a kind of 'oh yes' or 'of course' and seems to be something you *already know*, but are seeing 'for the first time'. This is something I've written about in the context of 'cosmic consciousness' and 'gnosis' in the mystical experiences of William James, R.M. Bucke, and P.D. Ouspensky.[27] Ouspensky, for example, had a tremendously powerful mystical experience looking at an ashtray – admittedly he was on nitrous oxide at the time – and William Blake, we know, saw heaven in a wildflower. Ouspensky had seen ashtrays before and Blake had seen flowers, but at the time of their mystical experiences, they were seeing these everyday things in a non-everyday way. The mystical experience, then, is not about *what* you see but *how* you see it.

The life world

The everyday world is the one we live in most of the time. It is the world of newspapers, television, and the internet. It's the world of politics, pop stars, and celebrities, of advertising, social networks, and sports figures. It is an almost exclusively human world, but mostly of the 'only human' variety. The philosopher Edmund Husserl (1859–1938) coined the term 'life world' to describe the particular environment in which an individual lives. I, as a writer, live in a particular 'life world', one different from, say, a bus driver's or a pastry cook's, although the three of us share a great deal of the larger, outside world. We may, for example, all take the same bus or eat at the same café or shop at the same market. Husserl borrowed the term 'life world' from the Baltic German biologist Jakob Johan von Uexküll, who spoke of an *Umwelt*. By *Umwelt* Uexküll meant the perceptual world of an organism. Uexküll describes, for example, the *Umwelt* of an amoeba, a jelly fish, or a sea urchin. An *Umwelt* (German for 'environment') is an organism's own particular 'world'. Husserl developed this idea into what he called the 'communal life world'. This is what we mean by the everyday world.

When millions of people watch the same television program or the same video that's gone 'viral' on the internet, they are all living in the same communal life world. When we read the papers or listen to the news on the radio, we are sharing a communal life world with millions of people we will never know personally. As our world becomes more and more global, this communal life world grows in size and as more and more people's lives are fed with content coming from the communal life world, there is a real concern that their lives will become little more than a reflection of the communal life world's values. Most of these are of a trivial character, as a look at newspapers, television, the internet, and other sources show. Increasingly we seem to be living in a communal life world that reflects almost exclusively the 'only human.'[28]

The other world is one outside this communal life world. It, too, is a human world, but it is more of the 'fully human' variety, which, by definition, means it extends beyond our immediate, personal concerns. Its essential difference is that, while we are in it, we see things *differently*. They take on a deeper, more significant meaning and intensity. Values are greater and seem to revolve around objective meanings, not the subjective, strictly personal, often trivial ones that inform the communal life world. (How important is it, really, to know the latest gossip about some celebrity?) In the other world we seem to be in touch with a reality far more powerful than the one we know in the communal life world. The simplest things, like a wildflower or an ashtray, can convey an entire universe of significance. Most of us live in the communal life world most of the time, and it has to be said that many of us are quite happy there and would feel disorientated anywhere else. But there are some who find it claustrophobic and suffocating and go to great lengths and often dangerous extremes to escape it. They are those Colin Wilson calls 'outsiders'. Where they want to get to is not always clear, but they all share the knowledge that, as seen from the other vantage point, the values of the communal life world are unimportant and ultimately diminishing.

Heimweh

Earlier I quoted E.T.A. Hoffmann's remark about music's ability to stimulate in us an 'inexpressible longing' and its relation to an 'unknown realm' quite separate from the 'outer sensual world'. What Hoffmann refers to is *Sehnsucht*, a term that, more than any other, expresses the essence of German Romanticism. There really is no strict English equivalent of *Sehnsucht*, which is defined by *Langenscheidt's German-English Dictionary* as 'longing, yearning, hankering, pining, languishing, nostalgia'. 'Unfulfillable' or 'inexpressible' longing – as above – comes closest, although for our purposes, 'nostalgia', with its sense of 'homesickness' is most apt. As I've written elsewhere, '*Sehnsucht* conjures up horn calls far off in the dark forest, the poignant glow of the sunset, which we will never reach, no matter how quickly we race to the horizon, the snow-capped peaks of a distant mountain range.' It speaks of 'beauty, distance, and the sense of something infinitely desirable just beyond our grasp'.[29] The German poet most associated with *Sehnsucht* is Georg Friedrich Phillipp von Hardenberg (1772–

1801), better known by his pen name Novalis, who captured the essence of 'unfulfillable longing' in the symbol of the elusive 'blue flower' from his unfinished novel *Heinrich von Ofterdingen*. Novalis, a deeply Hermetic thinker, felt *Sehnsucht* for most of his short life – he died at the age of twenty-eight – and he encapsulated the pull it had on his consciousness in his best known aphorism 'All philosophy is homesickness'. This is an insight with which many of the thinkers looked at in this book would agree, and this homesickness wasn't felt only by nineteenth century Romantics. In the 1960s and 70s, it sent quite a few twentieth century Romantics on the mystic trail. In *The Journey to the East* (1932) Hermann Hesse wrote about a group of travellers from all times and places, some real, some fictional, perennially on the move toward 'Home'; he even quotes Novalis: 'Where are we really going? Always home!'[30] Along with his other novels that became hugely popular during the Hesse 'revival' of the 60s and 70s, *The Journey to the East* became a favourite among the many who had felt a sense of homesickness in those decades, and who took to the road, looking for a cure for it

One thinker who would agree with Novalis was his older contemporary, the French theosophist Louis Claude de Saint-Martin, whose early life in many ways resembles Novalis' own.[31] Both were sickly youths who spent a great deal of time on their own.[32] For Saint-Martin, homesickness, or, in German, *Heimweh* is 'the nostalgic desire for the heavenly fatherland'. Man, Saint-Martin said, 'has a mind which is not satisfied by this world, not satisfied to ruminate upon dregs of matter', and which 'would be badly served by a wisdom which rests only in visual and material nature ...'[33] Like the Kabbalists, Saint-Martin believed that his mission was to remind man of his proper task, which he saw as that of minister of the creative word, as a collaborator with God in perfecting creation. Like Milosz, Saint-Martin believed that through some primal crime, man had lost his true place in the cosmos, and was now trapped in a fallen world. Saint-Martin expresses the 'inside out' character of our existence in his maxim that 'It is necessary to explain things by man and not man by things', which means that we should regard the cosmos as an expression of man – Adam Kadmon, the primal Man – and not man as a product of the cosmos, of the physical forces at work in it, which is, of course, how materialist science sees us. For Saint-Martin, our place in the cosmos is unique. As mentioned, it 'differs from that of other physical beings, for it is the reparation of the disorders in the universe'. This and other 'counsels of the exile' – a general title given to Saint-Martin's writings – place Saint-Martin clearly in the tradition of *tikkun*. It is unclear how versed in Kabbalah Saint-Martin may have been, if at all, and the figure of 'the Repairer' is for him associated with Christ. But as we've seen with the Hermetic teachings, the idea that man has a responsibility here, unlike any other created being, is not the property of one spiritual teaching, but seems to be an archetype shared by many. The Baal Shem Tov, the eighteenth century mystical rabbi and founder of Hasidic Judaism, said 'truth is always in exile'. As noted above, Saint-Martin called his collected writings 'the counsel of the exile'. This suggests that both, in different ways, shared the same insight about mankind.

We should point out that the Kabbalistic tradition has it own sense of *Heimweh* or *Sehnsucht*. *Reshimu* is the impression of the Divine Light left behind after the *Ein-Sof* contracts itself in the *tzimtzum* necessary for creation. In this sense, we can see *reshimu* as a spiritual equivalent of the 'cosmic background radiation' astrophysicists believe is conclusive evidence for the Big Bang. We can think of the cosmic background radiation as the faintest physical tremor, the furthest ripple away from the initial explosion. In the case of *reshimu*, the radiations, however, are holy. One poetic description of *reshimu* is that it is like the fragrance of wine left behind in a glass after it has been emptied. In ourselves, the *reshimu* manifests as a sense that we have forgotten something, and it also prompts us to look for what that is. A helpful analogy is that of a tune we can't quite remember. If some other tunes come to mind, it's the *reshimu* that tells us they aren't the one, rather like a inner music critic who makes sure we don't confuse muzak with the music of the spheres.[34]

This sense of having forgotten something is at the heart of the western philosophical tradition. All knowledge is recollection, Plato tells us. Martin Heidegger (1889–1976), whose task it was to bring western metaphysics, begun by Plato, to its end, lays the blame for our modern nihilism on our 'forgetfulness of being', an assessment he shared with Gurdjieff. Heidegger's 'fundamental ontology', his analysis of our 'being-in-the-world', centres on the idea of *Sorge*, which, for the existentialists who followed him, meant 'anxiety' but which in our context is better translated as 'care'. For Heidegger we are 'the shepherds of being', the trustees of the irreducible fact that we and the world *are*. Our loss of wonder at this miracle is the evidence of our forgetfulness, our 'fallenness' into what Heidegger calls the 'triviality of everydayness', his term for Husserl's 'communal life world'.[35] (The idea that the Higgs-Boson or any other elementary particle could 'solve' the secret of the universe would for Heidegger be but another manifestation of our amnesia.) But whether it is the *reshimu*, *Sehnsucht*, *Heimweh*, or what George Steiner called 'the nostalgia for the absolute' – 'We are the creatures of a great thirst,' Steiner says, 'bent on coming home to a place we have never known' – there is something in us that makes us dissatisfied with 'the world' and leads us to look for that elusive *something else*.[36]

What that is, however, may be closer to home than we think.

Two worlds, two brains

Earlier I mentioned the psychiatrist and philosopher Iain McGilchrist, and quoted from his book *The Master and His Emissary* (2009). In it he provides an enormous amount of evidence suggesting that our experience of living in two worlds has a neurological and anatomical basis. This is the fact that we have two brains, each of which perceives the world very differently. The thesis of McGilchrist's book is that 'for us as human beings there are two fundamentally opposed realities'.[37] One is the world presented to us by the left brain, the other is the world presented to us by the right brain. I would add to McGilchrist's two opposed realities a third, conciliatory one, which McGilchrist himself includes as his argument unfolds: the world presented to us when both brains work together. This rare and somewhat miraculous unity is, I suggest, at the heart of moments of insight, of mystical experience, poetry, creativity, and the peak experiences that Abraham Maslow argued come to psychological healthy people, the 'fully human' who achieve self-actualisation. It is a kind of 'Goldilocks moment', when, rather than being at odds with each other, the two opposing sides collaborate and achieve the 'just right' state of creative balance. They 'complete the partial mind', as W.B. Yeats argues happens in moments of crisis and intensity.[38] This collaboration between two opposed realities housed in the opposite sides of our brains is, I suggest, a neurological expression of the *coincidentia oppositorum*, or unity of the opposites, by which, according to the Kabbalistic, but also the alchemical, neo-Platonic, Hegelian, Jungian, and other systems of thought, the shattered universe is repaired.[39] There is a good argument, then, for the idea that both the fractured, shattered world we live in – *Sitra Achra*, the Other Side – and the means of healing it, are contained within our heads.

McGilchrist contends that what is important about the fact that we have two brains, or two cerebral hemispheres, is not *what* each hemisphere does, but *how* it does it. We are all familiar with the popular literature on the right and left side of the brain, how the left is a 'scientist', geared toward logic and language, and the right an 'artist', attuned to feeling and intuition. This neat separation, in which certain 'functions' are localised in one side of the brain or the other, is, however, misleading. By the 1990s, 'hard' neuroscience more or less abandoned the split-brain as a serious subject for study because ongoing research had discovered that, in fact, both sides of the brain are involved in practically everything we do. Both sides, it turned out, *do* the same things after all, although one side may be more involved in an activity than the other. So, rather than a scientist and an artist living in separate flats, having nothing to do with each other, the two actually share a two-bedroom bungalow, with each pretty much having the run of the place. This recognition, coupled with the fact that the 'scientist/artist' divide had been appropriated by a host of popular New Age and self help books, was enough for official neuroscience to back away from the split brain and turn itself to other matters. The fact that we have two brains turned out to be not important after all, it seemed, and some researchers even joked that we have a second brain as a kind of spare in case something goes wrong with the other one.

Yet McGilchrist contends that there is an importance difference between the hemispheres, but it is not, as mentioned, in *what* they do but in *how* they do it. In a nutshell, McGilchrist argues that the two brains each have a very particular 'take' or attitude toward the world, and that these are radically different from each other. The right brain presents us with reality as a unified whole, a *gestalt*. It gives the 'big picture' of a living, breathing, 'other' – whatever it is that exists outside our minds – with which it is in a reciprocal relationship, bringing that 'other' into being through our consciousness of it, while being itself altered through the encounter. It is essentially geared toward working with living things and with perceiving overall patterns and meanings. One myth about the split brain that McGilchrist punctures, is the idea that the left brain is 'dominant', while the right is a kind of helpful, but not essential, sidekick. The truth, he argues, is the exact opposite. The right brain, he says, is primary, it is the 'master', who provides us with our fundamental connection to reality. It is through it that we 'participate' with the world. The left, he tells us, is the right brain's 'emissary'. It's job is to manipulate that primal reality, to analyse it, and break up the whole into well-defined parts, that it can then control, and its basic way of doing this is by representing the 'other' in a form like itself. Unlike the right brain, the left feels more comfortable with mechanical things, with bits and pieces that it can take apart and put back together into something it is familiar with. Its relationship to the world is much more 'detached' than that of the right. While the right brain seeks out the new, the unfamiliar, and the living – whatever is present to it *now* – the left is geared toward what it already knows, and what it knows best, McGilchrist tells us, is the machine. We can say that one gives us a world to live in, and the other the means of surviving in it. And although we need both the big picture and the small detail in order to be fully alive and fully human (not 'fully functioning' as 'functions' are themselves an expression of the left brain's tendency to mechanise experience), the two opposed 'takes' often clash, as if we tried to look at a panorama and through a microscope at the same time. We need to see the forest *and* the tree, and this sometimes proves difficult. It would be impossible for one brain to do this simultaneously, and so we have two.[40]

    Throughout our history, the two brains have been engaged in a more or less friendly but serious rivalry, McGilchrist tells us, inhibiting and restraining each other's excesses in a neurological embodiment of Blake's dictum that 'Opposition is true friendship', and evening each other out in a process resembling the give-and-take of Hegel's dialectic. The right brain needs the left because its picture, while of the whole, is fuzzy and lacks precision; it sees the world, say, with the warm glow we enjoy after a glass of wine, when things take on a indefinite but strongly *felt* significance, which we would be hard pressed to put into words, which are the left brain's tools in trade. The 'emissary's' job, then, is to unpack the *gestalt* the right brain presents, and then return it to the right, increasing its definition and clarity.

The left needs the right because while it can focus on minute particulars, in doing so it loses touch with everything else and can easily find itself adrift. We can think of it as an obsessive who, fixated on some fine but negligible point, forgets why he was doing something in the first place, or as someone suffering from a perpetual tunnel vision, powerfully focused on a tiny detail, but blind to the surrounding field. A right brain imbalance would, say, provide the kind of consciousness that animals may have, a warm, general, but inarticulate sense of the great surround, in which we would be embedded, with no sense of detachment or distance. (Nietzsche once remarked that it was no good asking cows the secret of their happiness, as they forget the question before they can answer it.) A left brain imbalance would present a world made up of bits and pieces, a kind of jig-saw puzzle of independent parts, with which we would feel no connection and which would appear to us as lifeless, each bit having a crystalline clarity but without relation, other than mechanical, to any other. The left brain's insistence on clarity produces an exacting attitude of 'either/or', an insistence on something being either 'black' or 'white', 'true' or 'false', as opposed to the right brain's openness to metaphor and an attitude of 'both/and'. (The right brain, McGilchrist tells us, has a sense of humour while the left is adamantly literal.) A left brain imbalanced world would be a world, say, in which Sartre's 'nausea', when objects take on a strange, threatening foreignness, and seem disconnected from their context, would be the norm. This is the kind of world, McGilchrist argues, that schizophrenics live in, in which there is an acute separation between perception and feeling, and it is the kind of world, he believes, in which we increasingly live today.

What has happened in recent times – since, McGilchrist suggests, the Industrial Revolution – is that the left brain, the emissary, has usurped power from the master, the right brain, and has become, as mainstream neuroscience contends, dominant. This is a kind of neurological expression of the Gnostic myth of an idiot demiurge who believes he is the true god and within whose world we are trapped. This has happened in other periods, just as there have been periods when the two sides reached an exceptionally creative accord. McGilchrist suggests that pre-Socratic Greece, the Renaissance, and the Romantic movement were such times of creative 'progress through contraries', to paraphrase Blake. But although other left brain imbalances were evened out, our own modern period is different, McGilchrist argues, because since the Industrial Revolution the left brain has been busy, steadily building a 'world' that is like itself – modern cities, factories – with the result that there is less and less of any 'other' for the right brain to present.

We can say that for the last two centuries the left brain has been erecting its own 'communal life world' at the expense of the right, and this itself can account for the kind of low status the right brain has been accorded by mainstream neuroscientists. Most neuroscience is conducted under the aegis of 'scientific materialism', the belief that reality and everything in it can ultimately be explained in terms of little bits, whether atoms or genes. We recall Francis Crick's remarks in the last chapter. Yet materialism itself is a product of the left brain's tendency toward cutting up the whole into easily manipulated parts, so it is not surprising, then, that materialist-minded neuroscientists *would* see the left as boss and the right as second fiddle. It is also not surprising that in such a world we would find ourselves in a purposeless universe that emerged for no reason when less than nothing exploded, and whose infinite extension can be cut up into an infinity of parts. As the 'other' that the right brain 'presences', in McGilchrist's coinage, is increasingly fashioned to the requirements of the left brain – precision, clarity, definiteness, and of parts, not wholes – we will increasingly find ourselves inhabiting a self-enclosed reality, an hermetically sealed 'communal life world', from which it will be increasingly harder to escape, as increasingly there will be no 'where' to go. As McGilchrist makes clear, this left brain hegemony informs practically everything in our life: the arts, society, education, even sex and personal relationships.

The other world

The 'where' we could go to, would be that other world, that other reality, housed in the other side of our brain. And we already have an intuitive sense of a world that is much more of a whole, that is much more connected, that is much more alive and related to us in an immediate, organic way. It is a world we inhabited much more when we were children and which we have a Proustian sense of when, say, we listen to music or read poetry (right brain 'media') or simply relax on a summer's day. It is also a world much more amenable to the insights of imagination – again, in the sense that Blake and Swedenborg and poets like Kathleen Raine, who spoke of a 'learning of the Imagination', understand the term. The problem is that this other world is not immediately communicable in language, or at least not in the kind of language that the left, the usurping emissary, will accept, the factual prose of logic or quantitative measurement. (We live, as the literary philosopher Erich Heller remarked, in an 'age of prose'.)

These meanings are implicit, given, self-evident. They exist in and arise from what the philosopher Michael Polanyi called 'the tacit dimension' and are an expression of the truth that, as Polanyi remarks, 'we can know more than we can tell'. This is another example of the idea that the *content* of mystical, poetic, or spiritual experience is often no different from the content of our everyday (left-brain dominant) experience. What makes the difference is *how* that content is experienced. People experiencing that sense of wholeness which the right brain provides find it difficult to convey this experience to someone else, because all they have to do this with are the words the left brain uses to get about its daily business. Unless you have had the experience as well, these words will say nothing more to you than what they usually do, and you will have no idea what your friend is talking about. Or if the experience is conveyed, through poetry or philosophy, the left brain's reaction is 'That's all very nice but *this* is what's really important', and it will get back to work manipulating and controlling the world. McGilchrist cites philosophers like Hegel, Heidegger, and Wittgenstein, who all had an intuition of this and who all had immense difficulty expressing it clearly, because they are trying to do something with language which language – a predominately left brain tool for ordering the world – wasn't made to do. Hegel and Heidegger frequently tie themselves in dizzying dialectical and ontological knots, and Wittgenstein gave up, concluding that we cannot *say* the truth, only *show* it.

Two modes of consciousness

As I read *The Master and His Emissary* I became increasingly excited at the new avenues of thought McGilchrist was opening up, only a fraction of which I can touch on here.[41] One thing that excited me immensely was that the distinction between the different 'takes' or attitudes toward the world exhibited by the left and right brain was already familiar to me, through the work of other thinkers that I had written about in earlier books. In *A Secret History of Consciousness* and *The Quest for Hermes Trismegistus*, I discuss the ideas of the maverick Egyptologist René Schwaller de Lubicz (1887–1961). A central theme in Schwaller de Lubicz's work is the distinction between what he calls 'cerebral consciousness' and the 'intelligence of the heart'. 'Cerebral consciousness', which Schwaller de Lubicz associates with our own modern mentality, operates by 'granulating' reality, turning it into discreet bits and pieces, and severing the connections that run through it, like the fibres in a spider's web, the links that allow William Blake to see a world in a grain of sand and heaven in a wildflower. These links make possible what Colin Wilson has dubbed the 'relationality' of consciousness, its ability to spread out into a tapestry of meaning. If we think of consciousness as a pool into which ideas or impressions are dropped, and the ripples from them radiating out and lapping against other ideas and impressions, and creating further ripples, we will get a picture of what he means.[42]

'All in the universe,' Schwaller de Lubicz wrote, 'is in interdependent connection with all'.[43]

It is the 'intelligence of the heart', Schwaller de Lubicz argued, that allows us to perceive and feel this 'interdependent connection'. This is a 'participatory' mode of consciousness, of the kind related to how the right brain perceives the world, and which allows for us to have a living connection to it, unlike the left brain's detached, analytical stance. For Schwaller de Lubicz, the intelligence of the heart 'allows us, in love, to be the thing, to be inside the thing, to grow with the plant, to fly with the bird, to glide with the serpent ...' It allows us, as he writes in his gnomic work *Nature Word*, to 'tumble with the rock which falls from the mountain/Seek light and rejoice with the rosebud about to open' and 'expand in space with the ripening fruit'.[44] It is also a mode of consciousness that allows for what Schwaller de Lubicz calls 'the simultaneity of opposite states', an ability to perceive two different realities simultaneously, without the need to decide on one or the other being true, an idea Schwaller de Lubicz absorbed from his study of quantum physics, specifically Werner Heisenberg's 'uncertainty principle' and Niels Bohr's 'complementarity'. 'Complementarity' tells us that light can act both as a wave and a particle, which are mutually exclusive states, at least to our 'cerebral' consciousness, and the 'uncertainty principle' tells us that we can know only the speed or the position of an elementary particle, not both, again a situation our 'cerebral' consciousness, with its demand for exactitude and clarity, finds difficult. The right brain, as remarked above, has an attitude of 'both/and', which allows it to appreciate metaphor, the linguistic sleight of hand in which one thing stands for another, and which to the literal 'either/or' minded left brain is incomprehensible. This suggests a connection between Schwaller de Lubicz's ideas and McGilchrist's findings.

In my earlier work I remarked that Schwaller de Lubicz's ideas about these different modes of consciousness are similar to ideas about consciousness presented by the philosopher Henri Bergson (1859–1941), and I suggested that, as Schwaller de Lubicz had been a student of the painter Matisse, and Matisse himself was influenced by Bergson, the connection was natural. Bergson was an immensely influential philosopher in the early part of the last century, but sadly is little read today. Bergson argued that our brains and nervous system serve an essentially *eliminative* function; they are designed to keep stimuli and information *out* of consciousness, and to allow only as much of it into awareness as is necessary for us to survive; in other words, to ignore 99% of the 'interdependent connections' everything has with everything else. This makes perfect sense, as a consciousness constantly awash in an awareness of everything would prove unhelpful, to say the least. As I've remarked elsewhere, Jorge Luis Borges' story 'Funes the Memorious', in which a character is paralysed through such an overload of knowledge, makes this clear. This reduced, highly edited consciousness presents us with a solid, stable, three-dimensional 'world' that it has 'carved' out of what is really a continuous, ever-changing flow, the streaming flux of becoming. This edited picture allows us to survive admirably – it has, indeed, made us the dominant species on the planet – but the picture of the world it provides is by definition incomplete. What it shows us, Bergson argued, is the 'outside' of things, their surface. It cannot show us their 'insides'. In order to perceive this, we need a different form of consciousness, which Bergson called *intuition*. By intuition Bergson means 'the kind of intellectual sympathy by which one places oneself within an object in order to coincide with what is unique in it and consequently inexpressible'.

Analysis, our everyday form of consciousness, which I referred to in my earlier books as 'survival consciousness', is, Bergson says, 'the operation which reduces the object to elements already known, that is, to elements common to both to it and other objects'.[45] That the left brain is geared, McGilchrist tells us, to what it already knows – to the familiar – and that the right is geared toward the new and living –with which it has a reciprocal relationship – suggests to me that what Bergson and McGilchrist are talking about is the same thing.

Meaning perception

This insight into the character of our two modes of consciousness was discussed at length in a short book, *Symbolism: Its Meaning and Effect* (1927), by the Anglo-American philosopher Alfred North Whitehead (1861–1947), again, another brilliant mind little read today. Whitehead argued that we possess two distinct ways of perceiving the world, what he calls 'presentational immediacy' and 'causal efficacy', which, for ease of comprehension, we can translate to 'immediacy perception' and 'meaning perception'. Immediacy perception gives us the bare facts, the discreet things that populate our awareness, the surface of Bergson's analysis and the 'granulated' bits and pieces of Schwaller de Lubicz's 'cerebral' consciousness. Meaning perception is the glue that holds these separate items together to form a whole and allows them to make sense. It is a form of Schwaller de Lubicz's 'intelligence of the heart' and Bergson's 'intuition', which allows us to get into things, to know their 'insides'.

If we think of what happens when we read a book we can get an idea of how these two different ways of perceiving the world operate. You, now, are looking at this page. In each sentence you perceive individual words, and in those words individual letters. You do not, however, consciously add the letters together to make a word, or the words together to make a sentence, and, eventually, the sentences together to make this book. You perceive these things as wholes, and it is through this that you absorb the meaning I am trying – no doubt clumsily – to express. If you perceived these words or sentences solely with immediacy perception they would make no sense. The individual words would appear in sharp clarity and definiteness, but they would have no meaning – aside from their dictionary definition – which is a kind of aura that hovers around a sentence or book, but which could not be discovered by adding all the individual words together. We experience this if we are tired and try to read. We read a sentence or two and fail to 'grasp' its meaning, even when we re-read it again. This is because our meaning perception has broken down and is not linking the individual words – the bits and pieces – together into a whole. When Sartre's protagonist in *Nausea* is frightened by a door knob or the root of tree, it is because his meaning perception is failing to provide the context – the atmosphere – in which these things are usually perceived. Meaning perception is most apparent in our appreciation of music. We known Beethoven's String Quartet

Op. 132 means something – indeed something profound – but we would be hard pressed to pin this down exactly, because this meaning exceeds language's ability to express it in any specific way. Such a specific meaning would, indeed, seem totally inadequate, just as we would be hard pressed to say what the meaning of a sunset is, although we would *know* it was beautiful, moving, poignant.

Science, the form of immediacy perception par excellence, can *never* answer this question. It can never tell us why a sunset or a string quartet is beautiful. This is no argument against science, merely an acknowledgement of its limits. It can analyse the sunset into wavelengths of light and the effect of refraction on them as they pass through the earth's atmosphere, just as it can analyse Beethoven's music into vibrations in the air, which is what music is. But it will never arrive at how these purely physical phenomena can produce in a sensitive consciousness a mystical feeling of beauty and awe. (Indeed, in neuroscience, this is known as the problem of *qualia*, how 'qualitative' phenomena – colour, sound, beauty, awe – can arise from quantitative ones – neurons or molecules.) The same would be true if I could somehow run to the horizon and grab hold of the sunset – which, indeed, I often would like to do. Even if I had it in my hands, I could not find its meaning that way, because although I perceive the physical qualities that convey the beauty to me through my senses, the beauty itself is something beyond the senses. And again, it is meaning perception that gives to the content of mystical experiences their peculiar quality; it adds another dimension to what we normally perceive in the mode of immediacy. This is why we feel we are seeing something for the first time. We are, or at least we are seeing it with our meaning perception cranked up a bit.

Exact fake

In fact, Whitehead developed his ideas about our two ways of perceiving in response to what he called the 'bifurcation of nature', the splitting of reality into 'primary' and 'secondary' qualities, which began with Galileo (1564–1642). Primary qualities are everything that is amenable to scientific quantification and measurement, and, since Galileo, whatever meets this criteria has come to be seen as the 'really real'. Secondary qualities are everything that give *value* or *meaning* to our experience, such as colour, sound, smell. (This is the essence of the 'fact/value' divide.) What is really real in primary terms about a sunset are the wavelengths of light and the index of refraction, both measureable and quantifiable and amenable to being analysed into separate parts. What is meaningful or valuable to us about a sunset is its beauty and grandeur, neither of which can be quantified or broken down into constituent parts. What should be clear from all of this is that Whitehead's immediacy perception is a left brain affair, and that meaning perception is right brain business.[46]

Whitehead's agreement with McGilchrist, Bergson and Schwaller de Lubicz is clear in a statement he made in regard to 'the absurd trust' we place in the 'adequacy of our knowledge', left brain knowledge, that is. The exactness that scientific thought prides itself on, and which it pursues in its analysis of the universe is, Whitehead says, a 'fake'. We cannot arrive at 'an adequate description of finite fact' – the bits and pieces – because such a description relies on a 'background of presupposition which deifies analysis by reason of its infinitude' – in other words, its 'interdependent connection' with everything else. We understand how each fact fits into the context under examination through common sense, the implicit meanings understood by the right brain, which arise from Polanyi's 'tacit dimension'. But there can be no analysis of common sense, Whitehead tells us, because it 'involves our relation to the infinity of the universe'. To grasp reality – which is what philosophers aim to do – we cannot, Whitehead tells us, 'rely upon any adequate explicit analysis', on the crystal clear part separated from the vaguer whole. Logic, the application of a rigorous either/or, which provides these parts in matchless clarity – Schwaller de Lubicz's 'granulation' – is a 'superb instrument', but it needs a background of implicit self-evident truth – common sense – in which to work. 'The final outlook of Philosophic thought,' Whitehead concludes, 'cannot be based upon the exact statements which form the basis of special sciences.'[47]

Guardians of the world

It may seem that in speaking about the two sides of the brain and two modes of consciousness we have moved quite a bit away from trying to understand ourselves as caretakers of the cosmos. But in fact the connection is quite clear. And I should point out that McGilchrist and the other thinkers I have discussed in relation to his work are not saying that the left brain is bad and the right brain good, or that immediacy perception is a villain and meaning perception a hero. The point is not to favour one at the expense of the other, but to see where one mode has been developed and has gained a kind of superiority and acceptance in our experience, while the other has been relatively ignored, if not actively discouraged. In our time we have erred, I believe, on the side of immediacy, but too much meaning perception can be just as much a problem as too little. As happens, I believe, under the influence of psychedelic drugs, to let the 'taps of meaning' run full force often results in a flood. We need both sides of our brain as well as both modes of consciousness in order to be fully human. And we should not be too concerned about localising the two modes too rigidly in the two cerebral hemispheres. Although I believe there is a strong link between them, what is important is to recognise that we do have these two different ways of relating to reality and that they operate in us already. So whether or not they can be localised in one or the other brain is, in a sense, irrelevant. They are part of our experience. In fact, they are the way in which we *have* experience, and it behoves us to see this and understand its implications.

Yet, there is a link between how our brains work and our relationship to the cosmos. McGilchrist points out that while the left brain is geared toward *controlling* reality, gaining a mastery of it, and, more or less, subduing it to its purposes, the right brain is more geared toward *caring* for and about it. If the one provides us with the ability to be 'exploiters of the world and one another', to be detached manipulators of its bits and pieces, the other helps us to be 'citizens one with another and guardians of the world', to participate in its coming into being. And, as I hope to make clear further on, also in its on-going evolution.[48] Let us take a look now at how this might be done.

3. Doing the Good that You Know

In his book *Kabbalistic Metaphors*, Sanford L. Drob quotes the Talmudic scholar Rabbi Adin Steinsaltz who remarked that we live in 'the worst of all possible worlds in which there is yet hope.'[1] Readers familiar with the seventeenth century philosopher Gottfried Wilhelm Leibniz (1646–1716) will recall his belief that, contra Rabbi Steinsaltz, we live in the *best* of all possible worlds. Leibniz believed that God could not have created an inferior world, and Leibniz devoted his considerable logical powers to explaining why there is evil in the world, a theological effort known as *theodicy*, or 'justifying God's ways to man'. Leibniz's somewhat optimistic assessment was famously parodied by the French *philosophe* Voltaire in his comedic work *Candide* (1759). Voltaire was only too aware of the evil in the world, and his satiric character Dr Pangloss (his name says it all) is perpetually explaining why some horrible situation is actually the best that could be hoped for. In a sense, Rabbi Steinsaltz's remark is a combination of Leibniz and Voltaire. We could say that it brings together the best and worst of both possible worlds, the overly optimistic, and the overly pessimistic.

In an interview, Steinsaltz explained what he meant by his remark. 'If I want to test a new car, the way that I test it is not on the smoothest of roads, under the best conditions. To have a real road test ... I would have to put it under the worst conditions in which there is still hope. I cannot test it by driving it off a cliff, but I can test it on the roughest terrain where I must come to the edge of the cliff and have to stop'.

It may seem a homely analogy, but Steinsaltz's view of creation isn't far from his ideas about testing cars. Creation, he says, is 'an experiment in existence', an 'experiment in "conquering the utmost case"'. Like testing cars, creation, he believes, 'would have been pointless unless it was precisely under these difficult circumstances'.

Our world, Steinsaltz believes, 'is on the brink ... If it were to be slightly, just slightly, worse than it actually is, then its basic structure would become entirely hopeless'. But Steinsaltz does not believe this is a pessimistic view. 'If a person sees the world as all pink and glowing, he is not an optimist, he's just a plain fool. An optimist, on the other hand, is one who in spite of seeing the terrible facts as they are, believes that there can be an improvement'.[2] In the Kabbalistic view, this effort of improvement is the work of *tikkun*.

Why God, who is supposed to be all-powerful, all-knowing, and all-loving, would create a world in which evil is rife, in which life is difficult and full of suffering, is a question that has plagued man practically since he began to think. To many, the answer to this conundrum is that there is no God. To others, of a particularly rosy bent, the answer is that there is no evil or suffering; they are, they say, an illusion, a sentiment enjoyed by many New Agers today. Neither conclusion seems satisfying, at least to me. I see too much good in the world not to feel that there is *something* behind it – exactly what, is something I am trying to articulate by writing this book. But I also see too much evil, pain and suffering to accept that they are simply illusions, to be swept away by my enlightenment, or that they are kinks in the human machine, to be straightened out by science and pharmacology. Earlier I mentioned the Kabbalistic belief that in creating the world, God or the *Ein-Sof* shattered the cosmic vessels on purpose, that the mistakes made in the beginning were part of the plan, and that all along the secret aim of the Big Spill – hidden even to the *Ein-Sof* itself – was for man to arrive and put things straight.[3] All the king's horses and all the king's men may not have been able to put Humpty Dumpty together again, but as far as Kabbalah is concerned, man's *raison d'être* is precisely that. But by repairing the broken pieces of existence *tikkun* does not return the world to a previous perfect state. On the contrary, it is only through man's work in this fallen world that creation itself is complete, and that the *Ein-Sof*'s work is accomplished. In this sense, the Humpty Dumpty we put back together is a better version than the original. From this point of view, the *Ein-Sof*'s initial handiwork would have remained inferior if the vessels hadn't shattered and man hadn't been sent to deal with the mess. Here, the repairman does a better job than the creator.

That the difficulties in existence, its pain, suffering, and evil, are inflicted upon us not as a punishment for Adam's sin, or because of the warped mind of an idiot demiurge, but because it is only through our grappling with them that the work of creation can be fulfilled, that the meaning of existence can be revealed, is an idea shared by different schools of thought. They all centre around the recognition of the necessity of struggle, of effort, and the understanding that it is only through working in a difficult world that any true value can be achieved. A block of stone is a difficult medium to work with. It is obdurate and resistant, and requires a great deal of effort to get something out of it. But every sculptor knows he can do nothing with a cloud. Creation *needs* resistance, otherwise it can't 'land a punch'. A bird in flight may seem to move through a perfectly 'free' space, but its graceful motion depends on the pressure of its wings *against* the air. 'The world we live in,' Sanford Drob tells us, 'is absolutely necessary for the "completion" of God'. By being 'exiled' into a fallen world, Drob says, the *Ein-Sof*, in its form as humanity, must face the 'material, intellectual, spiritual, and moral adversity' that enables it to fully actualise the values at its centre.[4] 'Only in a material world of chaos, toil, and trouble', Drob tells us, can 'the values, which are mere abstractions in the heavens ... become fully real'.[5]

Every artist, every creator of any sort, knows this: that an idea is only as good as its expression. As long as they remain solely in the freer, ideal realm of the mind or imagination (in the negative sense), ideas, insights, and visions may retain a certain perfection, but they lack effectiveness. They can only acquire this if they come down from the ideal heights and enter the lowlands of reality, and this is not easy work. This is why many people who have a hankering to be creative never get around to actually doing it. They are afraid that their beautiful idea will lose its perfection in the frustrating attempt to capture it, and they are reluctant to make the effort to try. They prefer to keep their brilliant idea safe. But as long as it remains solely in their head, or is trotted out occasionally in conversation – we all know would-be writers who can talk a good book – it remains unreal. This is something we will return to in other thinkers we will encounter in this book: the need for spirit to embody itself in matter, for values to be actualised in life.

(In a sense we can say that creation is an attempt on the part of God to self-actualise, in Maslow's term.) The good, the true, and the beautiful remain pleasant abstractions until they are forced into the resistant medium of reality. There is no escaping this. Our business is to make values real and the only place we can do this – the only place it *can* be done – is here, now, in our fallen world. As the philosopher Max Scheler, who we will return to further on, remarked, 'An assimilation of the forms of being and value with the actual effective energies can take place only in the raging tempest of the world'.[6] Gurdjieff once told Ouspensky that the earth is in a very bad place in the universe, almost the worst, the equivalent of Outer Siberia.[7] Everything here is difficult and costs a great deal of effort. But it may be the only place where we can get things done.

Saving God

Scheler's remark may seem dramatic, but some actualisers of value have found themselves in 'raging tempests' that were more than metaphorical. The Cretan poet and novelist Nikos Kazantzakis – best known in English speaking countries for his novels *Zorba the Greek* (1952) and *The Last Temptation of Christ* (1960) – wrote his philosophical credo *The Saviours of God: Spiritual Exercises* (1960) in Berlin in 1923, during the collapse of the Weimar Republic, an economic cave-in that makes our recent 'credit crunch' seem like a picnic. Photographs from the time show people pushing wheelbarrows full of worthless currency in order to buy a loaf of bread. At one point a US dollar was worth 4,200,000,000,000 *Deutschmarks*, a figure more reminiscent of astronomical distances than rates of inflation.[8] Kazantzakis tells how before writing the work he was afflicted with a strange malady, that the Freudian psychologist Wilhelm Stekel, whom Kazantzakis had consulted in Vienna, explained to him was known as 'the saint's disease'. Kazantzakis wrote to his wife, telling her that he was confined to his bed, dosed in antiseptics, his face wrapped in compresses he had to change by the hour. He told her he was suffering from eczema, but this was not really true. His face had swollen to a 'loathsome blubber of flesh' and his lower lip oozed a peculiar yellow liquid.[9]

At night he would awaken, crying in pain. Any attempt at a social life, at attending a concert or lecture, only made things worse. All he could do was remain in his room and work on his verse drama about the Buddha and his renunciation of the flesh, a work he was to abandon and recast as his *Saviours of God*.[10]

Doctors he consulted could provide no help; they had no idea what was wrong with him. But Stekel offered a clue. He told Kazantzakis that he had 'a spiritual and mental energy beyond the normal' and that his body suffered from this. He also told Kazantzakis that his affliction was known to the monks of the Middle Ages, that it would come to ascetics who were on the brink of breaking their vows. 'When they found the temptations of the flesh too much to endure,' Stekel told the poet, 'they would run howling toward town for a woman. But on the way, to their great horror, their bodies broke out in sores and boils, their faces became flushed and bloated, a yellow liquid dripped from pores until they fell on their knees in repentance ...'[11]

Although he kept it from his wife, Kazantzakis knew why this medieval sickness had come to him. One evening in Vienna he had gone to the theatre and by chance found himself sitting next to a beautiful woman. His fiery, impassioned writings suggest the opposite, but Kazantzakis was a shy man, a paradoxical character trait he shared with one of his heroes, Nietzsche. (Nietzsche famously said that he was not a man but 'dynamite', yet he was too shy to propose to Lou-Andreas Salome, and had to ask a friend. Understandably, she rejected him.) Kazantzakis was surprised that within a short while he found himself talking to his enchanting neighbour. Bored with the play, they left together and spent the rest of the evening walking. To his surprise, Kazantzakis found himself inviting his new friend to his room. She excused herself, saying she could not come that evening, but would be happy to the next. Kazantzakis returned to his room, excited by his encounter and anticipating the next evening's delights. But when he awoke the next morning, his 'eczema' had broken out; his chin and lips had swollen enormously. He sent word to the mysterious woman, apologising for having to cancel their assignation and arranging it for the next night. But his condition worsened daily, and he had to keep putting her off. After days of suffering and solitude, his face wrapped in bandages, he went to the opera to relieve his boredom, and it was there that he met Stekel. In the end, Stekel explained that, as with the medieval monks, his affliction was caused by his breaking his vows, not to his wife, but to his spirit. 'Your body,' he told the suffering poet, 'is suffering from remorse of spirit'. Kazantzakis had determined to write his play about the Buddha and had become obsessed with the question of asceticism. Now this temptation threatened to interfere with this. Stekel told Kazantzakis that unless he renounced his 'lady of the theatre' and left town, his condition would only worsen. At first Kazantzakis could not accept this, and continued to see dermatologists. Eventually he took Stekel's advice. Almost as soon as he did, his 'eczema' was cured.

Understandably, Kazantzakis resented the suffering he endured, but within it he also felt a strange kind of joy. 'Because of my illness,' he wrote, 'my soul was filled with heroic bitterness. I understand those heroes now who worked amid bodily wretchedness. To endure bitterness, yes – but at the same time, out of pride, not to enlarge your bitterness but to reach the opposite extreme – to invoke joy and health as though they were the general law. Never before have I been so prepared to perform a valiant deed as in these days when I am filled with loathing as I look on my face ...' 'A mystical joy,' he wrote, 'penetrates my life ... because I am testing my endurance ...'[12]

Less than two months after this, Kazantzakis destroyed the three thousand verses he had written on the Buddha, and devoted himself to a new work, which, he said, would state the terms of his own salvation. This was *The Saviours of God*. It is a powerful declaration of his faith in man as the agent of 'creative evolution', something along the lines of Julian Huxley's 'transhumanism', which we encountered in Chapter 1, but which Kazantzakis approached in a more mystical manner. Kazantzakis was a believer in the philosopher Bergson's *élan vital*, or 'life force', which he saw at work in the world, creating God. 'If we are to set a Purpose,' he wrote, 'it is this: *to transubstantiate matter and to turn it into spirit*' (italics in original).[13] Kazantzakis realised he was a 'weak, ephemeral creature made of mud and dream'. But he also felt 'all the powers of the universe whirling within me,' a statement that has a striking similarity to the Hermetic idea of man being a creature of two worlds.[14] The essence of this work, of 'saving God', Kazantzakis tells us, is 'struggle'. 'My God', he says, 'is not almighty. He struggles, for he is in peril at every moment; he trembles and stumbles in every living thing' 'God is imperilled' and 'he cannot be saved unless we save him with our own struggle'. Contrary to the teachings of the established church, Kazantzakis asserts that 'It is not God who will save us – it is we who will save God ... by transmuting matter into spirit'. This salvaging operation is not easy and there are no guarantees: 'If we all desire it intensely, if we organise all the visible and invisible powers of earth and fling them upward, if we all battle together like fellow combatants eternally vigilant – then the Universe might possibly be saved'.[15]

These last remarks are reminiscent of how David Lindsay, who we met in the last chapter, views the 'muspel fire', symbol of the 'true' reality, that Maskull, the hero of his Gnostic novel *A Voyage to Arcturus*, seeks. 'Muspel,' Lindsay writes, 'was no all-powerful universe, tolerating from pure indifference the existence side-by-side with it of another false world, which had no right to be ... Muspel was fighting for its life.'[16] For Lindsay, like Kazantzakis and other thinkers encountered in this work, the 'muspel-fire' – or whatever you want to call the drive within us that compels us to go beyond ourselves and become something more – is not all-powerful or all-mighty but needs our help, an idea that the established church, as well as many 'alternative' spiritual teachings, not to mention materialist science, does not accept.

Responsibility

The struggle to save God, to transmute dense, resistant matter into living, creative spirit, produces, Kazantzakis tells us, 'two superior virtues'. They are 'responsibility and sacrifice'. 'Love responsibility,' Kazantzakis commands. 'Say: it is my duty, and my duty alone, to save the earth. If it is not saved, then I alone am to blame.'[17]

Another actualiser of values who also viewed responsibility as central to our work in the world was the psychologist Victor Frankl (1905–1997), originator of the school of 'logotherapy', a form of existential and humanist psychology which sees *meaning* – not sex, power, or self-esteem – as the 'primary motivational force in man'.[18] Frankl saw responsibility as so important and fundamental a virtue that he believed a Statue of Responsibility should be erected on the west coast of America, to balance out the Statue of Liberty that stands on the east. Freedom is, of course, important, but Frankl believed that it is 'in danger of degenerating into mere arbitrariness unless it is lived in terms of responsibleness'.[19] In a time when people's 'rights' are increasingly enlarged to include wider and wider areas of life – from the right to have the state provide us with a living and keep us healthy, to the right to have children, whether we are biologically capable of it or not – Frankl's idea may seem reactionary, but there are plans to make his statue a reality.[20] Whether or not the Statue of Responsibility gets off the ground – and if it does it is unclear if it will make a difference – Frankl's philosophy is clearly an example of Rabbi Steinsaltz's belief that values must be tested in 'the worst possible world in which there is yet hope.'

The actual world – life world – in which Frankl developed his ideas may have been one in which even Rabbi Steinsaltz's austere optimism would have been severely tested. Steinsaltz remarked that 'there are indeed worlds below ours in which there is no hope at all'.[21] In 1942, when Frankl, his wife, and his parents were sent to the Theresienstadt concentration camp, he could have been excused for believing he had entered such a world. Certainly by the time he was transferred, first to Auschwitz, then to Kaufering (associated with Dachau), and his wife was sent to Bergen-Belsen, where she died – his parents were murdered in Auschwitz's gas chambers and his brother died there too – he could not have been blamed for believing he had entered a hopeless world. In 1945, Frankl was liberated from the Türkheim camp by the American army, and practically the first thing he did was write his book, *Trotzdem Ja Zum Leben Sagen: Ein Psychologe Erlebt das Konzentrationslager* (1946) ('Saying Yes to Life In Spite of Everything: A Psychologist Experiences the Concentration Camps'). Originally published in English as *From Death Camp to Existentialism*, it later achieved world acclaim as *Man's Search for Meaning: An Introduction to Logotherapy* (1959). At the time Frankl died, the book had sold ten million copies and was translated into more than twenty languages.

If anyone knew about living in the worst of all possible worlds in which there is yet hope, Frankl did.

Given the conditions under which he lived for five years, it would have been easy and understandable for Frankl to have arrived at a philosophy of despair, of hopelessness, and to have developed a conviction that life was surely meaningless.[22] Many of his fellow inmates did precisely that. But Frankl discovered that precisely the opposite was true: that life retains meaning even under the worst possible conditions, and that suffering, rather than something we should avoid or deaden through drugs, 'may well be a human achievement'.[23] (By 'suffering' Frankl does not mean the pointless neurotic complaints many of us bring upon ourselves, nor does he mean we should seek out suffering in some masochistic self-flagellant way, but the real, 'existential' suffering that life unavoidably brings to us.) Frankl recognised that those inmates who possessed a sense of purpose, of some goal lying ahead in the future, were more likely to survive, and that their physical health was directly related to this as well. Without goals, without some purposeful anticipation, we live, Frankl said, only a 'provisional existence', a kind of marking time which is really a death in life. Yet what is important about this sense of anticipation isn't necessarily a definite goal set for a specific future date, but an attitude, a 'way of being', that sees the world as one in which goals, purposes, and meanings are worth pursuing and can be achieved. It is precisely the opposite of the scientific view of man as a 'lumbering robot' housing a collection of 'selfish genes' or a chance outcome in the 'Monte Carlo game of life' It is essentially a belief in free will and in a world in which meanings are real, not 'projected' by our own deluded subjective minds. It is an acceptance of ourselves as active agents, not passive stimulus response machines, acted upon by outside forces.

Frankl turns around the standard question 'what is the meaning of life', which we address to the universe, demanding it respond before we agree to participate in it. Instead, Frankl argues that 'each man is questioned by life; and he can only answer to life by *answering for* his own life; to life he can only respond by being responsible.' What is demanded of us, Frankl says, is not 'to endure the meaninglessness of life' as Jacques Monod, H.P. Lovecraft,

John Gray and others would have us do, 'but rather to bear [our] incapacity to grasp its unconditional meaningfulness in rational terms'. This is the kind of meaning we find in music, in sunsets, and in love, none of which are amenable to 'rational terms'. 'Thus,' Frankl wrote, 'logotherapy sees in responsibleness the very essence of human existence.'[24] There is meaning in everything, but it is up to us to draw it out.

Intentions

In Kabbalah the work of *tikkun* is carried out through what are called *kavannot*, or 'intentions'. Although there are specific acts, rituals and prayers – *mitzvoth* – required by the faith, *tikkun* itself can be performed while doing practically anything. What is important is not so much *what* you do as *how* you do it, a theme we have encountered a few times already. In fact, as Iain McGilchrist suggests, the kind of *attention* that we bring to the world, determines the kind of world we will encounter. 'Through the direction and nature of our attention, we prove ourselves to be partners in creation, both of the world and of ourselves'.[25] This is because 'the kind of attention we bring to bear on the world changes the nature of the world we attend to.'[26] The 'patient and detailed attention' that, according to McGilchrist, is characteristic of the right brain, will reveal a world very different from 'left brain attention', which is geared toward control and manipulation.[27] (We can say the same of Whitehead's meaning perception and immediacy perception or of Bergson's intuition and intellect or of Schwaller de Lubicz's intelligence of the heart and cerebral consciousness.) The kind of attention that sees a world in a grain of sand is different from the kind that sees the same grain of sand as a troublesome thing in your shoe. Performing *mitzvoth* without the appropriate *kavannot* means going through the motions of an empty ritual, and it doesn't take much reflection to recognise that most of our actions in the world lack the appropriate *kavannot* or attention. Most of our actions and encounters with others are really social lubricants, whose purpose is to make things move along smoothly. We ask someone how they are, but we don't really want to know, just as the cheery 'have a nice day' is repeated so many times at the checkout queue that it becomes a mindless mantra. Shop managers want their employees to give the appearance of having a solicitous concern for the customers' welfare, and of course some do honestly wish us a nice day. But much happy service is really routine, the act without the intention. It is empty, like the husks of the *klipoth*, or like the absurd 'thank you' we are given by automated 'self-serve' machines. A machine cannot thank us, it can only make the sounds of the words 'thank you', just as it makes other sounds. (And if we ever start replying 'you're welcome', or, more apt for our time, 'no problem', we will really be in trouble.)

Luckily, as Daniel C. Matt remarks, 'since all material existence is animated by the Divine, even the most mundane activity can serve as an opportunity to discover God'.[28] There is no special act of *tikkun* nor a special person entrusted with its pursuit. As Gershom Scholem explains 'it is not the act of the Messiah as executor of the *tikkun*, as a person entrusted with the specific function of redemption, that brings Redemption, but your actions and mine.'[29] Therefore it is once again up to us. And bringing attention to our acts, especially the little ones, has a double effect. In the first place having a true intention when you say 'thank you' – that is, *really* meaning it – can be felt, both by the person we address it to, and by ourselves. It is surprisingly refreshing to be 'real', as difficult as it may be in a huge megalopolis, filled with millions of people. It doesn't mean loving the other, in a Christian sense; we may never see the person again. But recognising that they *are* a person, and not an obstacle to be overcome, or a trouble to fend off, or a competitor to beat, reminds us of their reality and our own. But focusing on the little things, on the small stuff, effects a change in the quality of our consciousness too. In *The Journey to the East*, H. H. (Hesse's alter ego) remarks that 'I, whose calling was really only that of a violinist and story-teller, was responsible for the provision of music for our group, and I then discovered how a long time devoted to small details exalts us and increases our strength'.[30] A similar insight came to Samuel Butler (1835–1902), most known for his utopian fantasy *Erewhon* (1872), but who was also one of the first intellectuals to launch a concerted attack on Darwinian evolution, starting with his book *Life and Habit*

(1878). 'When fatigued,' Butler wrote, 'I find it rests me to write very slowly with attention to the formulation of each letter. I am often thus able to go on when I could not otherwise do so.'[31] Why should this be so? My own suggestion is that by making the effort of focussing, of consciously 'paying attention' (the phrase itself suggests the effort and value involved) we compel the left brain – or, if we want to avoid localising these distinctions in a specific part of our anatomy, our immediacy perception – to slow down, so that it begins to operate at the same speed as the right. Our immediacy perception begins to harmonise with our meaning perception and so both gives whatever we are focussing on more depth (an added dimension) and draws on reserves of power (strength) that we are usually not aware of.

The good that you know

Although he does not speak of *tikkun*, Emanuel Swedenborg captured its essence in his notion of 'doing the good that you know'. As the psychologist Wilson Van Dusen (1923–2005), a modern day exponent of Swedenborg's ideas, puts it, 'the Lord suffers a stillbirth until man acts by the good that he knows'.[32] As with Kazantzakis, God, or the Lord, needs to become real through human actions, and the most effective way of achieving this, for Swedenborg, is by doing the good that we know. But how do we know what is good? We live in a time when common sense is at a premium and hitherto shared ideas about goodness have become a rarity. Our moral and ethical sense has in many ways been eroded by notions of 'cultural relativism', which argue that the good is not an absolute, objective reality outside of ourselves – as all religions maintain – but a product of culture, that is, man-made. The old adage that 'one man's meat is another man's poison' has become a rule of thumb by which we avoid imposing our subjective notions of the good onto people who do not share them.[33] We do not, however, ask why doing this is good, if, that is, the belief that it is good to tolerate values we do not share is itself only a relative good? There may be cultures that do not value toleration. Should we be tolerant of them, or should we maintain that toleration of relative values is an absolute good, even if doing so undermines the notion of cultural relativism? Our obsession with rights has also obscured our sense of the good. It is common today to find almost daily a tabloid rant about how concern for his 'human rights' has led to a serious criminal being given a very light sentence, or for his case to be thrown out of court, while the victim or his family has in no way 'received justice'. Our sense of the good is also muted by a science that tells us that altruism and other supposed selfless values are really products of the evolutionary struggle for survival, and are not really altruistic or selfless at all. Or it explains that our supposed acts of choice, in which we choose to do what is right, are really an illusion, because we are unavoidably programmed to act the way we do.

In any case, the only kind of good that science or cultural relativism can point to is a utilitarian one, in which something is good if it helps us achieve some end, that is, if we can utilise it. So a hammer is good if I want to bang in a nail, but neither it nor anything else is good in itself. One wants to ask why banging in the nail is good. If we say 'to build a house' we are still left with asking 'to what end'? 'In order to live'. 'Yes, but to what end'? And so on until ultimately we arrive at some good for which we cannot posit a further utility and must simply accept as good 'because'. Utilitarian ideas about the good, then, ultimately rest on some good that is not utilitarian. Nietzsche remarks somewhere that confusion over ideas of good and evil is a sign that a culture is in decline and these sorts of conundrum suggest he may be right.

In such an atmosphere of uncertainty and purely functional thinking, it may seem that doing the good that we know will not be particularly easy. Yet one of the characteristics of self-actualisers, Maslow argued, is that they have 'far less doubt about right and wrong than average people do,' and self-actualisers, more than anyone else, will be concerned about *tikkun* and 'doing the good that we know'. It is as if, Maslow said, self-actualisers 'were perceiving something real and extra human'.[34] This is an expression of Maslow's discovery that self-actualisers 'are better cognisers and perceivers' and are 'more acute about differentiating fine hue differences', which makes them 'better perceivers of reality'.[35] It is not, as a politically correct critic might suggest, that self-actualisers have no qualms about imposing their ideas of right and wrong on other people, although, to be sure, some of the most horrific events in history – the Reign of Terror following the French Revolution, for example, or Stalin's forced collectivisation of farms – came about through the best intentions, through people doing what they believed was good. This is not the case with self-actualisers for the simple reason that self-actualisers are not interested in other people, but in their own struggle to develop. What it means is that self-actualisers are less likely to get confused about these things because they are more in touch with their own feelings and instincts – perhaps with the intelligence of the heart that Schwaller de Lubicz speaks of – and are less susceptible to the diffidence that comes with such confusion. Which is itself another version of the 'false humility' which Saint-Martin believed enfeebled us.

Uses, means and ends

Swedenborg himself had no doubt that the good is extra-human, that its origin is in the spiritual realms that are above the physical and which inform it, giving it meaning. The good has, at heart, what Van Dusen calls a 'general impulse toward existence'; it manifests as a necessity to actualise itself and an insistence on being embodied. As with Victor Frankl and Rabbi Steinsaltz, Swedenborg's good is made of values that require human action to be made real, and this making real begins, not with metaphysical or psychological or sociological ideas about the good, but with acts, whose simplicity and obviousness remove them from all the confusion of our moral dilemmas. Doing the good that you know does not require doleful soul-searching or philosophical sophistication. It does not ask 'what is good?' Instead it tells us that if we know a good, we should do it, and all of us know some good. This is part of Swedenborg's concept of 'uses'. 'The good that is in us,' Van Dusen writes, 'is nothing until it comes into existence as uses.'[36] These uses are as simple as sweeping the floor, washing the dishes, or doing the laundry. In fact, they can mean doing any job, however tedious, as long as it is done well and for its own sake. As the sociologist Richard Sennett makes clear, this is the essence of craft, 'the desire to do a job well for its own sake.'[37] As Maslow wrote, 'practically any means-activity [utilitarian] can be transformed into an end-activity ... Even the dullest, dreariest job, as long as it is worthwhile in principle, can be sanctified..'[38] While we can perform these tasks mechanically and some good will have been achieved – a clean floor, dishes, and clothes are not to be sniffed at – they can also be performed with *kavannot*, the kind of attention that brings their reality to life and also increases our own, as Samuel Butler and Hermann Hesse discovered. The Divine Love wants to be real, Swedenborg tells us, and it is made so through our actions. 'The proof that we are regenerated', Saint-Martin said, 'is that we regenerate everything around us.'

Swedenborg also recognised that our actions and efforts require difficult conditions – Max Scheler's 'raging tempest' – for them to be of value. Just as Jean Paul Sartre did, Swedenborg recognised that freedom is terrifying. But it is a necessary terror, because it is only through struggling with it that we can find our own way. 'Only what we accomplish with our own freedom,' Van Dusen writes, 'lasts to eternity'.[39] As Kazantzakis, Rabbi Steinsaltz, Victor Frankl and others recognised, 'it was not,' Van Dusen tells us, 'meant that existence would be all smooth and simple, like being bottle fed on divine pablum'. We are not, as many New Age and other spiritual teachings tell us, supposed to 'go with the flow'. As Victor Frankl learned under the worst possible conditions, it is through the trials and tribulations that life sends us, that we learn what is most meaningful. If everything was handed to us, we would never find out. Our path is that of most achievement, not least resistance. Our own existence, here and now, is the door to everything else, Swedenborg taught. We do not enter the spiritual realms through any other portal. This makes human reality once again the 'centre' of the universe, and with it comes a cosmic responsibility. It is through us that reality grows. The physical universe may be expanding, as scientists tell us. But the inner universe is also getting bigger. The 'saved', for Swedenborg, experience reality enlarging, getting bigger, opening up to the good that is larger than themselves. They participate in life's effort to discover and act on its highest meanings and possibilities. In effect, they help life do the good that it knows. By contrast the 'damned' live in an increasingly shrinking world, an existence that constricts by the day. If the 'general drift of the hierarchy of heaven is toward joining with others, toward the unitary oneness of the Lord', the 'drift of hell is toward separation, division, cutting off ...'[40] The saved are those who act by the good that they know, and they formed for Swedenborg the Church of the New Jerusalem. This need not be an actual physical building or an actual congregation. All that is required for membership is that one acts from the good that one knows. And if we look sincerely enough, we can find this almost anywhere. In fact, one good that we know, is to look for the good, so we may know it.

Sex

It may seem odd to bring sex into a discussion about *tikkun* and doing the good that you know. But for Swedenborg and Kabbalah, as well as for other spiritual thinkers and teachings, sex is a central means of repairing the universe. Needless to say, it is one good that practically all of us will have precious little hesitation about doing. In fact, we will more than likely go out of our way in search of it.

I believe we live in one of the most highly-sexed times in history, certainly in the modern period. Clearly people have always had sex; we would not be here if they hadn't. But sex in our time has in many ways become the sole unquestioned good; in the absence of any spiritual values, this isn't surprising. Certainly money, possessions and power are still highly prized. But sex seems to be the one uniform answer to our unhappiness. If we are feeling low and depressed, then we should have an affair. That will make us feel alive. And it also seems to be everywhere, with little of the modesty and intimacy that at an earlier time made sex mysterious and powerful, even transformative. Our electronic media keeps it pretty much on tap; it is, in fact, difficult to avoid. Because of this, if we are modest about it, and not 'open' or 'transparent', we are, according to contemporary lights, 'in denial', 'repressed', or 'neurotic', certainly not healthy. We should loosen up and relax. Viagra, and other 'performance enhancements' ensure that we can have a potent, commendable sex life, even into old age. The number of orgasms we have, as well as those we can elicit in a partner, are a sure gauge as to whether or not we are sexually well. Once again, quantity, not quality, is the key point, so go for it.

But as with everything else, sex without *kavannot*, without the attention required to make something real, is simply another empty act, a hollow husk, a coital *klipoth* which, although many of us will deny it, leaves us feeling unfulfilled, certainly not with the ultimate satisfaction that having a vigorous, even athletic sex-life is supposed to provide. Or so we are told.

Postmodern sex – if I can coin an ugly phrase – is perhaps most baldly expressed in the work of the French novelist Michel Houellebecq. In novels like *Atomized* (1998), *Lanzarote* (2002), and others, Houellebecq presents sex in our time as a commodity, a product of capitalism, totally goal-oriented and utilitarian, obsessed with performance and available only to those who can afford it. Sex tourism features centrally in his work, as do internet sex chat-lines and other forms of cyber-eroticism, which require no contact between people at all. Love, tenderness, and the vulnerability which used to be an element in erotic encounters have evaporated in this world of empty coital coupling, of vigorous but vapid 'shagging'. Houellebecq's deadpan descriptions of sex acts have been called pornographic in their detail and intent, yet ultimately they are not really sexy and are more like clinical or technical reports than the kind of literature 'one reads with one hand', as the novelist Angela Lambert said of pornography. Some characters don't 'do penetration' but are into 'oral'. Others devote themselves to other specialities or simply allow their bodies to be used, indifferent to being an object. Houellebecq's graphic descriptions of mechanical sex, combined with his characters' lack of emotional investment – the English title of his debut novel, *Whatever* (1998), says it all – give his work an air of misanthropy reminiscent of John Gray. In this regard it is curious that Houellebecq is a great reader of H.P. Lovcecraft, who also looked at human bodies as machines.[41] Houellebecq's early essay *H.P. Lovecraft: Against the World, Against Life* (written in 1991, translated in 2005) is a late contribution to the French appreciation of Lovecraft's nihilism. It is a vehement form of literary criticism, but its thesis that Lovecraft's work forms a critique of modernity ('Absolute hatred of the world in general, aggravated by an aversion for the modern world in particular. This summarises Lovecraft's attitude fairly accurately') was anticipated by Colin Wilson's *The Strength to Dream* (1962) by decades.

Cosmic sex

Against the kind of consumer, functional sex depicted in Houellebecq's novels (which, to my mind, skirt microscopically close to affirming the kind of inhuman eroticism they are ostensibly supposed to be criticising), sex as an element of *tikkun* presents a very different picture. For Kabbalah, 'the entire cosmic order is interpreted in sexual erotic terms.'[42] As mentioned earlier, when the *sephiroth* containing the divine energies shattered, the primal androgynous Man, Adam Kadmon, was split in two, fractured into separate halves, male and female, who forever seek their other self, in order to heal the rift. This is a motif widely shared; Plato used it in the *Symposium*. Loving eroticism between man and wife is a central theme in Kabbalah; it represents on the earthly plane the creative act of the Divine, and unites the male and female energies, polarised by the Big Spill. In Kabbalah, God is forever seeking to reunite with his Shekhinah, his female better half. The Tree of Life itself is polarised into two opposing pillars, that of Severity and Mercy, whose integration results in the balanced Middle Pillar, where the *sephiroth* Tiphareth (Beauty) holds a central place. Whenever a man and woman come together in loving union, this reintegration is achieved, and a part of the universe is repaired.

This reparation continues, even in heaven, if we are to believe Swedenborg. In one of his last books, *Conjugial Love* (1768), written in his eighties, Swedenborg tells us that angels have sex lives too, which, by his account, are much better and more fulfilling that ours on earth. In heaven angels are in 'continual potency' – a bad break for Viagra investors – and there is no weariness after lovemaking, nor post-coital sadness. Rather a cheerfulness of mind and eagerness of life. In heaven, married angels 'pass the night in each other's bosom', and the wife receives unreservedly 'the virile sentiments of the husband'.[43] This form of 'doing the good that you know' may have come to Swedenborg via Kabbalah.

During a stay in London in 1745, Swedenborg came into contact with the Moravian Brethren, followers of Count Zinzendorf, who included in their worship meditation on a kind of mystical eroticism focused on the passion of Christ.[44] Some of the Moravian worship included elements of the doctrines of Sabbatai Zevi, the 'false Messiah', which also included sex. During this time, Swedenborg also came into contact with the mysterious Rabbi Samuel Jacob Hayyim Falk. Falk was a Kabbalist from Galicia (Poland) who incorporated aspects of Sabbatian practice in his teachings. Falk ran an 'esoteric school' from a house in London's East End, and it is here that Swedenborg met Falk. It is possible that it is through Falk that Swedenborg developed his ideas about angelic eroticism. Swedenborg himself was highly sexed, and in his writings he advocates, in certain circumstances, keeping a mistress, and pre- and extramarital relations, in order to avoid 'the damage that can be caused and effected by too strict a repression of sexual love in the case of those who are troubled with a superabundance of sexual drive.'[45] Most Swedenborgians avoid the thought, but a superabundance of sexual drive was apparently a problem for Swedenborg.

The meaning of love

For *tikkun*, it is the erotic union, and not the product of it, that is important. This is to say that the ecstasy of the *coincidentia oppositorum*, and not the outcome of it, is what matters. The ecstasy is the sign that broken halves of the cosmos have been united. And again, the ecstasy is not simply a matter of having a good orgasm, or a sufficient number of them. Ecstasy is not the same as pleasure. Sex in human life is not a function, like digestion. It is at the core of our being. The ecstasy is not simply a physical reaction, like sending some powerful solvent through us, to 'clean out the pipes'. It involves parts of ourselves that we are 'normally' unaware of; I put 'normally' in quotation marks because our normal is subnormal for the 'fully human'. This is why Nikolai Berdyaev, who we met in Chapter 2, can argue that 'It is quite possible to say that man is a sexual being, but we cannot say that man is a food-digesting being'.⁴⁶ Obviously sex involves physical organs, just as digestion does. But it is not limited to them; it reaches beyond them to permeate our entire life. (This is why online advertisements promising 'increased performance' and 'better tools' – i.e. larger genitals – miss the point.) Utilitarian views about sex focus on its product, on reproducing ourselves, or on 'improving' the race through breeding. This is as much a functional view of it as that depicted in Houellebecq's novels. Against this practical view of sex, mystics and spiritual thinkers argue that what is important in sex is the change in consciousness that accompanies it. In his book *The Meaning of Love* (1892), the Russian mystical philosopher Vladimir Solovyov (1853–1900), an older contemporary of Berdyaev, argued against regarding sex as essentially a means of reproduction or improving the race. These utilitarian views reduced what was essentially a cosmic phenomenon to mundane, ignoble ends. As in Kabbalah, Solovyov saw the *task* of love as the work of creating 'an absolute ideal personality out of two finite ones, to *integrate opposites*.'⁴⁷ To become, that is, 'fully human' or 'actualised'.

Love – by which Solovyov meant sexual love – is, he said, the 'visible restoration of the Divine image in the material world, the beginning of the embodiment of true ideal humanity'.⁴⁸ This 'restoration' has, according to Solovyov, no other means to perfect itself apart from us, and our own regeneration, for Solovyov 'is indissolubly bound up with the regeneration of the universe and with the transfiguration of its forms of space and time,' something Saint-Martin knew when he said that the proof of our regeneration is that we regenerate everything around us.⁴⁹ We cannot achieve our full individuality in a vacuum. It can only be accomplished within 'the corresponding development of the life of the universe', which – in a statement that anticipates the anthropic cosmological principle by a century – 'from time immemorial has been interested in the preservation, development, and perpetuation of all that is really necessary and desirable for us'.⁵⁰ 'We are just as necessary to the world as the world is to us,' Solovyov tells us. And the 'failure to recognise [our] absolute significance is equivalent to a denial of human worth,' which, according to Solovyov, is 'the basic error and the origin of all unbelief'.⁵¹

Coming home

As we've seen, the significance of sex in human and cosmic life did not escape Solovyov's countryman Berdyaev. His chapter on sex in *The Meaning of the Creative Act* is one of the most powerful statements on the cosmic status of the erotic ever written. 'Sex,' Berdyaev tells us, 'is a cosmic force and may be comprehended only in its cosmic aspect'.[52] 'Not only in man alone, but in the cosmos as well, there is the sexual division into male and female and their sexual union'.[53] As many thinkers, poets, and mystics have felt, Berdyaev sees the feminine as the more cosmic side of the divide, with the masculine energies being predominantly 'anthropological', that is, human. Much rhetoric has been exchanged over the 'sexist' bias of such remarks. Yet while I can appreciate that, being a man, my views may be biased – and fully realising the criticisms I am leaving myself open to – I can only say that both my experience and my study confirms Berdyaev's view. Man relates to the cosmos, to nature, through woman – at least I do. Man, as a sexual being, is a *part* of nature, of the cosmos, while woman, *for* man at least, *is* nature, *is* the cosmos. Just as, and in the same context, sex is a part of man's being, but woman *is* sex.[54]

Which is not the same as saying that woman is only sex, or has no role or part to play in things except for a sexual one. Yet the notion of woman as cosmos is a dim, obscure insight into the significance woman has for man, in his dim, obscure hunger for cosmic union. This polarity is also, Berdyaev tells us, at the heart of the 'sex war', the bitter, sometimes ferocious antipathy that falls between the sexes. Just as a dim memory of our primal androgyny attracts us to our opposite, the primal catastrophe that ripped the original unity in two pushes us apart. It is this tension, this 'love hate', which gives relationships between the sexes their peculiar atmosphere. It is also why, for all the macho, predatory bluster about performance and potency, man finds a strange sense of 'home' in woman, and why woman is such a powerful trigger for *Sehnsucht* in men. (It is the 'eternal feminine' after all, that 'draws us onward', as Goethe recognised long ago.) Feminists may reject the idealisation of women by men, seeing in it really only another form of dominance. But I think they miss the point. Very little of the cosmos, I think, could be repaired under a strictly feminist rule, about as much as would be under a strictly 'masculinist' one, although in both there could still be plenty of sex. Under either, sex would remain purely functional, aimed at reproduction, pleasure, or as part of a 'healthy' regimen, giving a new meaning, perhaps, to our current dictum of 'five-a-day'. When a man admits it, and when he is not pressured by 'performance anxiety', and is not fixated on having an orgasm – which he can easily do, and with perhaps greater efficiency, by masturbating – he knows there is some sense of 'return' when he makes love (which is not the same as 'having sex') with a woman.

Phenomenological love

Berdyaev's idea that sex is a form of cosmic union was shared by Max Scheler. Scheler is little read today, but he was enormously influential in the early years of the last century, his wide ranging ideas on ethics, consciousness, art, sociology, and many other subjects informing the work of Martin Heidegger, Maurice Merleau-Ponty, Ortega y Gasset, Scheler's student the Catholic philosopher and saint Edith Stein, and Pope John Paul II, to name a few of his readers. At the centre of his work was his exploration of the nature of love, which he saw as not only a human emotion, but as a means of inquiry into the fundamental essence of being.

Love, for Scheler, was the *sine qua non* of phenomenology, which, in its essential form, is a way of allowing the world to be what it is, without interference by human concepts or aims. It is, in a sense, a way of listening to what the world has to say to us, from which follows the recognition that it has something to communicate, and is not simply a vast, inanimate machine. As a student of phenomenology soon discovers, however, there are as many phenomenologies as there as phenomenologists. Scheler's own approach differed from phenomenology's founder, Edmund Husserl, on a central point. Dismayed by the undermining of philosophy by variety of developments, mostly in science and psychology, Husserl wanted to secure philosophy as a rigorous discipline by arriving at a strict methodology. Scheler, however, saw phenomenology as more of an 'attitude' or 'way of seeing', than a method. It was a way of 'being with things' that would allow for them to show themselves as they are. It was a *Geisteshaltung*, a 'disposition of spirit', or 'spiritual stance', that one took toward the world, as one waited for it to show itself. In this sense, Scheler's philosophical approach shares much with Iain McGilchrist's suggestion that the kind of attention we bring to the world determines the kind of world we will attend to.[55] If love between people creates an atmosphere in which one can 'be oneself', then love directed at things will allow them to be themselves as well. After all, philosophy is the 'love of wisdom'. Through love we can know things that we would be unable to know without it; it has a real cognitive purchase. It is not only an emotion we feel about things; it is a way – *the* way, for Scheler – for us to know them as they are. Love opens the world to us, as it opens ourselves to the world, whereas hate, for Scheler, was a closing off, a shrinking away from things (it is unclear if Scheler ever read Swedenborg, but this seems very similar to Swedenborg's distinction between the 'saved' and the 'damned' mentioned earlier).

Scheler believed that it was not enough to apply a strict methodology to our experience in order to understand its structure. Without the phenomenological love that he speaks of, we would not have the experience itself. In this sense, philosophy is not a detached, objective observation, but an *encounter*. Again, this love is not showy or expressive or overtly demonstrative; we do not do phenomenology as Scheler sees it by hugging a tree. But it does participate in a 'patient and detailed attention to the world' – which is more akin to poetry – rather than an aggressive attempt to 'uncover' its secrets, which we associate with materialist science.

Unity with others

Through his investigation into the nature of sympathy, the bond of feeling we have with others, Scheler saw that 'man's sense of unity with the living cosmos is in general so bound up with the sense of union in sexual love that the latter is, as it were, the "gateway" to the former'. It is, he believed, 'the means, prescribed by Nature herself, of *arousing* in man a capacity for identification with the cosmos ...'[56] With Berdyaev and Solovyov, Scheler was critical of the idea that the purpose of sexual love was reproduction. He castigated the 'subordination of the values of life to those of utility', and 'the monstrous delusion that man can 'produce' men at his own will and pleasure (like so many cardboard boxes or machines) ...'[57] Sexual love for Scheler, as for Berdyaev and Solovyov, was essentially *creative*, and was not satisfied with mere reproduction. 'The qualitative peak of organic life' – sexual love – 'cannot be subordinated to the mere quantitative multiplication of human lives' he wrote.[58] 'True sexual love,' for Scheler, 'is a creative life-force, the nobility of life eternally blazing the trail upwards and outwards from its present level towards a higher form of existence'.[59] (Berdyaev agreed; he speaks of love as 'the source of an upward movement of personality' a 'creative upsurge into another world'.[60])

What this love creates are values, and deeper perceptions of meaning. Scheler in fact turns around the standard utilitarian view of the value of sex as a means of reproduction. He argues that, rather than see the transformative power of sexual love in the individual as an unintended by-product of a necessary means of continuing the race, reproduction itself should be seen as a good *because* it provides the individual with the opportunity of this transformation. As he writes, 'Rather than allow that this love exists simply for the sake of reproduction, we should proclaim it already entailed in the vital meaning and value of reproduction itself, that as many human beings as possible should discover the crowning-point of their lives in the experience of sexual love.'[61] Sexual love, then, isn't good because it gives us life. Rather, life is good because it gives us sexual love and the opportunity to experience the 'qualitative peak of organic life'. Scheler himself, it seems, enjoyed this qualitative peak fairly often. As his biographer John Raphael Staude points out, Scheler had a reputation as an eroticist, among other things, and that 'whenever he had to choose between his devotion to a transcendent God and the warm body of a woman, with pangs of conscience but unable to do otherwise, he chose the woman.'[62]

This sense of unity and the fundamental fact of love as a means of experiencing reality informed Scheler's ideas about society.

One problem that troubled Scheler was how we know other people's minds, what in the phenomenological literature is called 'intersubjectivity', a conundrum that troubled Husserl in his last years, and gave rise to his notion of the 'life world'. We have direct sensory experience of other people's bodies, but not of their minds. Generally this does not trouble us; we take it for granted that, just as we have an 'inner world', metaphorically located inside our heads, so too do other people. But if you stop and think about just how we know this, things become a bit less obvious. After all, as we know from dozens of science fiction films, a robot or android could be made so life-like that we could believe that it was human, just like ourselves, yet it would be just a machine. In fact, some forms of schizophrenia manifest in precisely this way, in a belief that people are not really people, but incredibly life-like robots. People suffering from this delusion have lost whatever it is that allows the rest of us to accept that those around us are human, just as we are, although we cannot 'prove' this in any 'scientific' way.

Scheler came to the conclusion that we have a direct perception of other selves, just as we have of our own. We do not infer that someone is feeling pain because their actions are like those we associate with pain. According to Scheler, we actually see their pain in their tears. We do not, as it were, observe the other, taking note of his behaviour, then tally up the various actions and come to the conclusion that, 'He is in pain', the 'bits and pieces' approach. We intuit it. In Bergson's sense, we 'get inside' the other, or, actually, are there to begin with. Scheler believed that originally, there was no difference between our inner world and that of others. Or more precisely, originally there was no such distinction between 'inner' and 'outer' or 'self' and 'other'. As Herbert Spiegelberg writes, in his monumental study *The Phenomenological Movement* (1976), for Scheler 'originally our social consciousness contains only a neutral stream of experiences, not yet assigned to either ourselves or to others'. 'The self and other', for Scheler, 'are discovered only as a result of a process of differentiation in the neutral primordial stream'.[63] (Which is another way of saying, as I do in Chapter 2, that our 'I' develops out of an initial 'ouroboric' unity with the world.) So rather than having to explain how my separate consciousness, located in my head, or more precisely, my brain, can know that you, too, have a similar separate consciousness housed in your brain – which is more or less the standard scientific view – Scheler argues that our sense of separateness is the result of a differentiation of an original primal unity. We are, in the deepest sense, parts of each other. If we are going to repair the universe, we won't be doing it alone.

4. The Good Society

One of the discoveries I made while researching this book is that the work of Max Scheler and that of Abraham Maslow have some very central themes in common. Although Scheler scholars and Maslovian psychologists will be quick to point out subtle important differences, Scheler's concept of a 'hierarchy of values' and Maslow's notion of a 'hierarchy of needs' are so similar to each other that, for all practical purposes, we can consider them, if not identical, then certainly informed by the same insight into human nature and reality. I don't know if Maslow was aware of Scheler's work, but I doubt it. Scheler's name doesn't show up in Maslow's books, and in the 1940s, when Maslow was laying the foundations of his psychology, interest in Scheler's work suffered in the United States because of a critical article forging a doubtful association between Scheler's ideas and National Socialism.[1] Scheler was not alone in suffering this fate; at the time, in America and Britain, Nietzsche, Hegel, and other German philosophers were all mistakenly tarred with the Nazis' brush, and required a subsequent critical rehabilitation.[2] In Scheler's case this was doubly ironic, as he was one of the first German intellectuals to warn of the coming fascism and his books were among the first that the Nazis banned. Because of this, Scheler suffered from an obscurity in Germany even greater than that in America.[3] Scheler, of course, could not have known of Maslow's work, as Scheler died in 1928, before Maslow had published anything.

What excites me about this sort of serendipity, which I've experienced with other thinkers, is the suggestion that what each thinker is focussed on is real, that is, it is an objective discovery about reality, not a subjective opinion or point of view.[4] This sort of independent corroboration is needed, especially when dealing with spiritual matters, which elude the kind of sensory or experimental confirmation enjoyed by positivist science. Two scientists dealing with a problem in molecular biology can perform identical experiments and receive confirmation of their theories there, 'before their eyes'. Philosophers and psychologists studying man's inner world or the world of meaning, do not have this option, and when two independent thinkers, separated by time, language, nationality, and the barriers of war, seem to be talking about the same thing, there is good reason to suspect that it is indeed just that, the same. And this suggests that it is real.

Just as Scheler believed that we have a direct perception of another's mind he also believed that we have a direct perception of values, that we see them in the same way that we see colours – not, however, with our eyes, but with our mind.[5] I don't decide that something is green, or yellow, or blue. I see that it is. The green, yellow, or blue is there, in the world, and I recognise it. In the same way, I don't impose a value on things, through a subjective act of valuation, providing what is really a neutral substrate with a spurious coating of significance or meaning, which is how a Lovecraft, Sartre, or Jacques Monod sees it. The value is there and I see it, just as I see the colour. (In the case of a sunset, mentioned in Chapter 2, the colour and the value seem interfused, as Wordsworth knew.[6]) 'There is a mode of perception,' Scheler writes, 'whose objects are totally beyond the grasp of the intellect, and for which the intellect is as blind as the ear and the source of hearing are for colour – a mode of perception nonetheless, which presents to us real objects and an eternal order among them – namely the values and their hierarchy'.[7] 'Values are not only valuations but also value-facts,' Scheler's interpreter, Werner Stark, tells us, 'that can be seen by our mental eyes in the same way in which our physical eye sees coloured surfaces'.[8] This insight led Scheler to speak about a 'realm of values' that we can enter, an objective, really existing world, like the physical world, that we can explore with our minds.

What was important about this insight for Scheler is that it provided an objective grounding for our values, for what we hold important and significant. And this was important because of the rise of the moral, ethical and other kinds of 'relativisms' that, as mentioned in the previous chapter, are so much a part of our modern and post-modern world. By declaring the world of values and meaning unreal – that is, not susceptible to quantification or measurement – science, or a particularly aggressive form of reductionist scientism, opened the door on the 'anything goes' school of values. Beauty, and everything else, was now in the eye of the beholder, not in any kind of objective stratum of real meaning or value. Like many other thinkers of the time, Scheler recognised that this was a dangerous development, which led, as mentioned earlier, to utility being the sole gauge of any good.

The hierarchy of values

Scheler saw values ordered in a hierarchy, reaching from the immediate but fleeting goods of the senses to the universal and eternal goods of the sacred. At the lowest rung of this 'ladder of values' is sensual pleasure, whether something feels or tastes good to us, is agreeable or disagreeable, whether we derive some immediate enjoyment from it or not, and whether or not it is of any use to us in achieving this enjoyment. This is the level of values that most of us are occupied with most of the time, and it is the level of values that informs utilitarian notions of the good and the 'consumer consciousness' that derives from them. At this level, our values are basically self-centred. One difference between the bottom and top rungs of Scheler's hierarchy is that sensual values are, as mentioned, transitory. We only feel pleasure as long as that which provides it lasts. Once we finish the chocolate bar or cigarette, the value it contains vanishes. Sacred values, however, continue long after what initially triggered them is gone. The force of a mystical or religious experience continues to be felt in our lives even though we may no longer be in immediate contact with whatever triggered it. Another difference is that sensual values are atomistic, meaning they can be divided into parts and shared. I can share a bottle of wine with a friend, and my half of the wine provides the same degree of sensual enjoyment as his. If I drank the whole bottle myself, I wouldn't get an increase in enjoyment, just more of the same. Sacred values, however, are holistic and cannot be divided. I can't have half a mystical experience, nor can I share it with someone else. It is an essentially private, inner experience, although the meaning of the experience can be shared. This is precisely what happens in religions, when the meaning of one person's experience – Christ's passion, Buddha's enlightenment – becomes the source of a way of life for others.

Above the sensual Scheler sees what he calls vital values, levels or degrees of life. These values inform the community or society we live in. One's life can be noble, heroic, and self-sacrificing, or it can be base, cowardly, and self-seeking. It can be healthy and vigorous, or sickly and decadent. One can do the good that one knows, or one can avoid it. One's relationship with others is informed by which of these is true. Positive vital values build a strong sense of community and social responsibility, while negative ones undermine our social and communal bonds. One of Scheler's concerns was that with the rise of utilitarian notions of value, which he believed were on the increase in the modern age, ideas of nobility, heroism, and self-sacrifice would lose their significance, and an ethos of self-seeking, in which everyone was out for themselves, would dominate. As altruism today is increasingly explained in terms of the 'cunning' of 'selfish genes', we can see that Scheler's concern was warranted.

Next in Scheler's hierarchy are intellectual or spiritual values. These are spiritual in the sense of the German *Geist*, the things of the spirit, or mind. Broadly speaking this is culture. These values are concerned with things that are goods or ends in themselves, the Platonic goods that inform the variety of different things that embody them. A landscape, an orchid, and a Rembrandt are all different but they are also all beautiful, and it is the beauty they share, rather than their particular expression of it, that is the concern of this rung on Scheler's ladder of values. These values are the opposite of utilitarian values, in which things are good only insofar as they help me to achieve an end. I do not listen to Bach's *B Minor Mass* with any end in view, nor do I appreciate beautiful scenery because of the potential value of the land. I listen to Bach and enjoy a mountain view because by cultivating my appreciation of them, I become 'more', my experience of being increases.

We may listen to music in order to relax; indeed, there are numerous collections of 'light classics' aimed at helping us do precisely that. But this would be a utilitarian appropriation of what is really a non-utilitarian, or spiritual good. One of Scheler's *bêtes noires* was precisely the invasion of the cultural world by the utilitarian sensibility, which promoted the cultivation of the arts and literature, provided they served some purpose, that they helped business in some way, much as meditation and other spiritual practices are today promoted as a means of increasing productivity. And while science, with its emphasis on 'know-how' and discovering the way in which things work, in order to be able to manipulate them, is an outgrowth of the lowest level of values, philosophy seeks to know things in themselves, their essences. It is driven by a sense of wonder, not a need to control and use. The value it embodies is again the opposite of the utilitarian. Trying to grasp the essence of being will in no way help me to control, manipulate, or in any other way use a particular being. My own being, though, can deepen and become more mysterious to me by doing so. This is an example of what George Steiner calls pursuing 'the sovereignly useless', an expression of a drive unique to humans: 'to be interested in something for its own enigmatic sake,' which, Steiner argues, 'may be the best excuse there is for man'.[9]

At the top of his hierarchy, Scheler places the sacred, the values of the holy. These values are involved in one's salvation. They answer questions about the ultimate meaning of existence and one's place in it. Here, the wonder and curiosity of the spiritual/cultural level give way to feelings of awe and reverence for the source of all being. And whereas at the spiritual/cultural level one can still maintain a certain detachment ,at this highest rung on the ladder of values, one is inescapably oneself, that unique personality, encountering the ground of its being. And it is precisely through this encounter that one is saved.

Fill in the blanks

Maslow was not a philosopher, but a psychologist and he spoke not of a metaphysical realm of values, but of human needs, evidence for which he believed he discovered in the human psyche. He was, in this sense, as empirical as any scientist, a claim he shared with his fellow psychologist Jung, who also believed he observed characteristics of the human psyche as demonstrably evident as the behaviour of elementary particles. In Jung's case, these empirical psychological facts were the archetypes of the collective unconscious, the expression and understanding of which, Jung believed, led to psychological maturity through a process he called 'individuation', how we 'become who we are'. In Maslow's case, they were what he called our 'hierarchy of needs', the satisfaction of which led to self-actualisation. I won't explore the connections here, but as any reader of Jung and Maslow soon discovers, individuation and self-actualisation are in all essentials practically identical.

Maslow believed that human beings have a set of psychological needs, of psychic requirements, that are as biologically inherent as our need for oxygen. They are, Maslow said, 'instinctoid', meaning we come supplied with them from the start. This notion of a kind of psychological blueprint for human beings, a set of innate goals that define what it is to be human, is out of sympathy with the tone of our time, which is more inclined to see nurture – the environment – rather than nature, as the key to human behaviour. This is, in fact, a dogma of modern political and social theory. All egalitarian politics is based on this idea, that at bottom, there is no inherent difference between one person and another, and any differences that do exist, are the result of outside, external forces – forces which, of course, man can learn to control and order.

Philosophers as different as John Locke (1632–1704) and Jean Paul Sartre argued, in different ways, that man's inner world is essentially empty, that it is like an unfurnished room, waiting to be filled up with things from Ikea. For Locke, when we are born we are *tabula rasa*, a blank slate, until our psyches are written on by experiences brought to us via the senses. Nothing is in the mind, Locke said, that was not first in the senses. Our inner flat is empty, until our senses deliver the furniture. For Sartre, as we've seen, man has existence but no essence; we are, for him, a kind of hole in things, and the project of human freedom – a negative state for Sartre – is the process of filling this hole with our choices. There is no human nature for Sartre, only a human condition. It is within human power, however, to change conditions, and we do so through our choices. But as Sartre denies any meaning to human existence, these choices are ultimately arbitrary. After he gave a rousing speech about human freedom and our responsibility to exercise it, Sartre was asked by some students for advice about what exactly they should *do* with their freedom. 'Whatever you like' was his reply. Not surprisingly, his students were not quite inspired by this.

Maslow disagreed with this 'empty' view of ourselves. He agreed with Sartre that we are free, but this freedom is not an emptiness. We are not free to make of ourselves whatever we want, forming ourselves out of whole cloth. Nor are we blank slates, waiting for experience to write on us. Although in different ways both Locke and Sartre champion human freedom, they also contribute to the idea that man can be made into anything, provided the requisite conditioning is applied, a doctrine that informed the Behavioural school of psychology. John Watson, Behaviourism's founder, claimed that as it was not susceptible to measurement and quantification, consciousness didn't exist. At least he had never seen it, and given this, it should be left out of any scientific account of human psychology. All we can look to, Watson said, was behaviour, and this insight led to seeing human beings as stimulus-response robots, flesh and blood vending machines which, supplied with the necessary coin (stimulus) would provide the required item (response). Neither Locke nor Sartre desired this conclusion, but their 'blank' vision of human psychology supports it nonetheless.

The hierarchy of needs

Maslow disagreed with Sartre and argued that there is indeed a human nature, which we ignore at our peril. Maslow's hierarchy of needs parallels Scheler's hierarchy of values in some very key respects.[10] One very important observation shared by Scheler and Maslow is that 'higher' needs or values do not appear until after 'lower' ones are satisfied. This is not to say that a higher need or value cannot take precedence over a lower one. This is the essence of any kind of self-disciple, and is behind the virtue of 'delayed gratification'. In order to pursue the spiritual/cultural value of writing this book, I must put aside sensual values (eating and drinking too much), vital ones (going to a party with friends) and even other cultural ones (listening to music). If I am starving or dying of thirst, I can't write at all, but once those values are satisfied, then the lure of the higher ones appears, or at least it can.

People who remain at the pleasure level of the hierarchy of values, as well as those whose highest values are vital, aimed at social goods, seem inadequate, precisely because they have not moved on to a higher value, once the lower one has been achieved.

At the bottom of Maslow's hierarchy are the basic needs for food, drink, and shelter. We all need to eat, we all need to drink, and we all need somewhere to sleep. These are the most primitive needs, shared by practically all living things; even plants need a bit of soil in which to grow. Above these needs is the need for security, for some emotional connection, for love, for some sense of belonging, for family of some kind. A starving man may not care about a roof over his head, but after he has been fed, the desire for some kind of home comes to him. Likewise, a homeless person may believe that all his problems would be over if he could only find a little perch somewhere, a small room he can call his own. But when that need is met, the desire for company, for a companion, arises, and he looks for someone to share his home.

If we meet these needs a new need arises, the need for self-esteem, the need for our worth to be recognised by others, by our friends, or workmates. My love need can be met by one other person or by a family, but my self-esteem need requires a larger group. So my workmates, the gang at the pub, the football team I play with on the weekend: these are the necessary sources of my self-esteem, my sense of worth, and I, in turn, serve the same function for them. Solitary types such as writers may receive self-esteem at a distance, from their readers and from reviews of their work (all too rarely, alas) but this is not quite the same as receiving it from others immediately around you. Sartre believed that most of us get our sense of ourselves, of who we are, from other people. This is one insight with which Maslow agreed, at least as it applied to the lower levels of his hierarchy.

Meta-needs

Maslow saw that the first three levels of his hierarchy were concerned with what he called 'deficiency needs'. They are all needs based on something we lack. I need food and drink to live. I need shelter for protection and some sense of physical security. I need love for emotional security and to avoid loneliness. And I need self-esteem to give me a sense of worth, of identity, and the feeling that I 'matter'. And without too much trouble, and allowing for a certain overlap, we can see how Maslow's deficiency needs parallel the lower rungs on Scheler's hierarchy of values. The pleasure values seem to equate with Maslow's most basic needs, for food and drink, and to some degree with the need for shelter and love, or more appropriately, sex. The vital values seem to parallel Maslow's needs for emotional security and self-esteem. When we reach the spiritual/cultural level of values, the pursuit of things which are ends in themselves, on Maslow's hierarchy we pass beyond the deficiency needs and move into what Maslow called 'being needs', and what I refer to here as 'meta-needs'. These meta-needs are inherent in each of us and each of us can, potentially at least, actualise them, just as each of us can pursue the spiritual/cultural value of being interested in something for its own sake. And yet, not all of us do.

All of us die if we don't eat or drink. We are all exposed to the elements and feel physically insecure if we don't have a home of some kind. We all feel a lack of human warmth if we don't have someone to love and who loves us. And we all feel insignificant and lack an identity if we do not possess the recognition of our peers. But not all of us feel that something is missing in our lives if we don't pursue the spiritual/cultural values. In fact, only a small number of us do. In Maslow's terms, that small number have passed beyond merely fulfilling their deficiency needs, to discovering an entirely different kind of need. This isn't a need based on something we lack, but one that arises from a desire to focus our energies on something more than satisfying our deficiency needs. And while we can see that the first three rungs of Maslow's hierarchy all fall under the umbrella of utility, in which our actions and choices are based on 'getting something', this is not true of these being needs. These needs are not aimed at getting or doing something 'useful', something that will help me meet the demand of the lower needs, but at things that are ends in themselves. In fact, meeting these being or meta-needs often involves saying no to our deficiency needs, just as pursuing spiritual/cultural values often involves postponing those of pleasure and life.

These being needs are essentially creative and are aimed, not at acquiring something we don't have, but at precisely the opposite: at giving something of ourselves. They express an overflow of our own being. It is at this level of Maslow's hierarchy that we enter the sphere of self-actualisation, that we begin to become, in his term, 'fully human', and hence able to perform *tikkun* and to take care of the cosmos. While we are primarily occupied with 'getting and spending', with fulfilling our collection of wants and desires, we are in some sense 'less human' and are concerned solely with taking care of ourselves. When we recognise that animals occupy only the lower rungs of Maslow's ladder of needs — those for sustenance, shelter, and some form of social life (but of course not all animals belong to groups) — we can see what this means. We are only fully human when we pass beyond these, as the being or meta-needs that lie ahead can be pursued only by us, or by beings like us. There is little evidence to suggest that animals can be interested in something for its own sake, can feel what George Steiner calls the 'speculative lust' for the 'drug of truth'.[11] What animal is concerned with truth at all? We speak of the curiosity of a cat, but this is not the same as the curiosity of someone fascinated with, say, styles of decorative ornamentation in early Egyptian pottery, and who will pursue this interest, often to the detriment of his 'creature (animal) comforts'. As far as we know, no animal wonders why it exists. Or, to put it another way, we are the only animals that do, and that wonder is precisely the threshold between our being only animals and being fully human. Whoever asked the first question about his existence was, by this reckoning, the first human.

This is not to speak ill of animals, but to recognise our difference from them. In our 'biocentrically correct' times, this is something we are loath to do, but it is, I believe, necessary so that we can grasp our responsibility to the cosmos. H.G. Wells (1866–1946), who we will return to shortly, made this point when he remarked that, for most of history, human beings have been 'up against it', focussed on fulfilling the demands of Maslow's lower needs (although of course Wells didn't use Maslow's terminology). During most of our existence we have been occupied with getting enough to eat, a place to live, a mate, and some sense of social rank, even within the primitive, proto-human hordes. But, Wells says, in recent times – the last few centuries – something new had entered the picture. A new kind of person had emerged. Wells calls him the 'originative intellectual worker', and with him came new demands, that have nothing to do with the deficiency needs that had driven human life so far. Now we can ask someone a question that would have made no sense in earlier times. 'Yes,' Wells says, 'you earn a living, you support a family, you love and hate, but – *what do you do?*[12] Wells recognised that fulfilling our deficiency needs was no longer enough, at least for some of us, who had moved beyond them into an entirely new territory. He recognised that precisely because as a species, we have been so adept at meeting our deficiency needs, we now have a surplus of time and energy to devote to 'originative intellectual work', work not driven by the utilitarian aims that occupied us for millennia. Hence eating, mating, and social bonding were no longer enough to define us as humans. A new kind of human had emerged, one that was *pulled* toward values rather than *pushed* by needs.[13]

Setting the norm

Maslow shared with Scheler his concern with values, but where for Scheler they existed in a metaphysical reality, for Maslow they were rooted in our very biology. Both, however, saw the need for a normative, not relativistic, approach to values, which means that for both values were objectively real and set a standard which we could either meet or fall below – set, that is, a 'norm'. For both there were criteria for deciding how well individuals met those standards, which were as real as anything materialist science could require. 'The question of a normative biology cannot be escaped or avoided', Maslow wrote. 'The value-free, value-neutral, value-avoiding model of science,' he said, 'is quite unsuitable for the scientific study of life.'[14]

Such a 'value-free' approach wanted to observe human life as if it were a thing, like an elementary particle or a chemical reaction, an expression of physical laws and necessities, or, at best, a collection of drives and appetites. But as Maslow, Scheler, and many other thinkers knew, such an approach distorts the very thing that it wants to study. Humans, Maslow recognised, were not only physical objects, stimulus-response robots, or a set of appetites, but free, creative beings with consciousness, aims, purposes, and the will to pursue them. If you omitted these characteristics or actively negated them – as Behaviourism did and much of contemporary cognitive science does – then, indeed, you could study people as if they were things, but whatever it was you were doing, it had little to do with human psychology, as you had excised everything human from your study. One virtue of this approach, however, is that if you treated everyone as if they were a stimulus-response robot or a mere physical object, with no regard for the values, meanings, or purposes that informed their lives, then clearly the question of whether or not someone failed to meet a norm or rose above it, would never arise, as there would be no norm by which to judge this. (Behaviourism saw people as basically machines, and there are, it is true, machines that work and those that don't. But machines that don't work we call broken, we don't say they have failed to meet the norm.)

After all, we don't set norms for how electrons should behave or for how iron should oxidise. We merely observe how these processes take place. The stimulus-response or object approach could apply equally in all cases, in that of a sinner or of a saint, in that of a child molester or a philanthropist. It didn't matter whether the person you were studying achieved something of value with their life or simply wasted it, as their behaviour, whatever it may be, was, from this perspective, simply the result of the physical forces or stimulus applied to them. Values, meanings, purposes, aims, were all nice things to think about, lovely stories we told ourselves, but in reality – at least for this approach – they were illusions. What was real was that when you did *this* to a person, he reacted in *that* way, and this would be true whether the person in question was a genius or an idiot, Mother Teresa or Hitler.

What was different about Maslow's approach to human psychology was that, unlike his predecessors, Maslow wasn't interested in what was wrong with human beings, but in what was right with them. He wanted to study psychological health, not sickness. Maslow in fact started out as a Freudian but eventually rejected Freud because he grew tired of studying damaged people. It is remarkable but when you think about it, up until Maslow began his work, practically all ideas about human psychology were based on material derived from psychologically ill people, from neurotics, psychotics, schizophrenics, and other forms of damaged psyches. This is rather as if all our ideas about health were based solely on studying sickness, and that health itself was defined as the absence of disease. But we all know that some people are healthier – more vital, more alive – than others, even if none of them is suffering from any specific illness. If we applied the same scheme to business, this would be like saying that if we break even with an investment, then we were doing well. But there is no idea of a profit here, only of not having a loss, and any business run on these lines would soon seem pointless.

Maslow grew disgusted with this scenario and decided instead to study healthy people. What he discovered is that, once people satisfy their lower, deficiency needs, they enter areas of human reality that range beyond our standard ideas of health. In fact with Maslow we arrive at a new idea of psychological health, the 'fully human'. Anything that falls below this we can consider as unhealthy, or at least as seriously lacking in some essential nutrients.

Eupsychia

One question that profoundly troubled Maslow in his last years was that of 'the good society', another concern that he shared with Scheler, who devoted a great deal of his energy to the problems of sociology. Maslow's writings on the peak experience have led some critics to see him as concerned solely with the individual. Indeed, the notion of self-actualisation seems to suggest this, and Maslow's work has been seen by some critics as an expression of a negative side of the 'human potential movement', as a hedonistic focus on the development of the self at the expense of others. A similar criticism has been made against Jung's concept of individuation. Martin Buber famously remarked that Jung was interested in an 'I-Me' relationship, not an 'I-Thou'. Maslow and Jung were concerned with the problems of society, yet both recognised that a society is only as good as the people within it.[15] Political movements aimed at changing the social structure would fail, they both saw, if the people within society did not change, that is, develop themselves. Yet both saw that self-actualisation and individuation can be severely hampered by a society that placed no value on these efforts. 'The actualisation of the highest human potentials is possible,' Maslow wrote, 'only under "good conditions". Good human beings will generally need a good society in which to grow.'[16]

Although both Maslow and Jung recognised that self-actualisation and individuation can take place under adverse conditions – indeed, in some cases adverse conditions positively foster it – both also recognised that the situation the world faced required more individuated and self-actualised people than the 'natural method' –for sake of a better term – could produce. In his last years Maslow devoted himself to the question of how best the process could be helped along.

'The good society,' Maslow wrote, 'is defined as 'that society ... which fosters the fullest development of human potentials, the fullest degree of humanness.'[17] If it is fair to equate being 'fully human' with embracing the responsibilities that come with this, those of *tikkun* and of 'taking care of the cosmos', then the good society will be one in which the values associated with these obligations will be promoted.

Clearly the society in which most of us live now does not. We live in a society that, for the most part, promotes exactly the opposite. We live in a society that promotes not the 'fully human,' but the 'only human'. Society as it is, our 'life world' as it is does not foster what is necessary for us in order to achieve our full humanity. This is clear. We are working at a handicap. As society and the world are, from the evolutionary point of view, still 'works in progress', this is understandable, and can only be expected in a 'worst possible world in which there is yet hope'. Nevertheless, it is still a problem. Maslow recognised this, and, like other social visionaries, he speculated on what a society that did foster our full humanity would be like. He called such a community Eupsychia, a coinage of Maslow's meaning 'the good psyche' or 'the good soul'.

If, as Maslow believed, 'the Good Person can equally be called the self-evolving person, the responsible ... fully human, self-actualising person', then the good society would also be informed with these values. The two go hand in hand; a feedback operates between them. Each needs the other. Maslow agreed with Julian Huxley that humanity had reached the point where it was responsible for its own evolution. 'We have become self-evolvers,' he wrote. 'Evolution means selection and therefore choosing and deciding, and this means valuing'[18] A self-evolving society will have to emphasise those values that promote its evolution. Maslow's Eupsychia is a place where that happens. Maslow's ideas led to important innovations in the workplace, as expressed in his classic work *Eupsychian Management* (1965).[19] His ideas about a 'eupsychian society' and on the anthropologist Ruth Benedict's notion of 'synergy' can be found in the posthumous collection of essays *The Farther Reaches of Human Nature* (1971). Anyone who reads these will find little to fault Maslow with, and it is difficult to disagree with him, that a society based on these lines would be preferable to one that ignored them.

Yet, although Maslow said that 'by Good Society I mean ultimately one species, one world', he also recognised that such a society would, at least in its beginning, be a 'subculture'. It would, at the start, be made up of 'only psychologically healthy or mature or self-actualising people and their families'.[20] This suggests that, after all, some of us can self-actualise, can become 'fully human', outside of a eupsychian society, otherwise where would these people come from? (Maslow himself grew up under very difficult conditions.[21]) This is a tension that runs through Maslow's late work: whether, given the proper conditions, society as a whole can self-actualise, or whether there is an inherent distinction between self-actualisers and those who do not reach the level of being needs. Why is it, Maslow asked, that more people do not self-actualise, do not pursue being needs, when their deficiency needs are met? Is it because of the way society is structured, or is it because of some inherent difference in people?

Or Utopia?

Earlier I remarked that H.G. Wells believed that in modern times, a different kind of person was emerging, one not satisfied with the occupations that had previously constituted human life. But

Wells too was troubled by the question of whether this was a sign of a general development in humanity, or the branching out of a new kind of human, a new species even.[22] In a late science-fiction novel, *Star Begotten* (1937), Wells even suggests that Martians are bombarding the earth with cosmic rays, in order to alter human evolution.[23] Wells himself was a firm believer in social progress and the possibility of improving humanity by creating a better society, and like Maslow, he envisioned his own kind of Utopia. After achieving fame with his early science fiction novels, *The Time Machine* (1895) and *The Island of Doctor Moreau* (1896),which present a curiously pessimistic view of mankind, Wells spent the rest of his career envisioning ways in which society and mankind could be improved. Practically everything he wrote in the twentieth century dealt with this question, but the essence of Wells' vision can be found in his first 'future society' work, *A Modern Utopia* (1905).

In *A Modern Utopia* Wells develops an idea that will obsess him for the rest of his life, that of the World State. It is a global government that has jettisoned all the petty squabbling that characterised politics in Wells' day, as in our own. The world is run by intelligent, efficient men, who are dedicated to promoting the common good. Nationalism and its wars are a thing of the past. The competition bred by capitalism is abolished, and a scientific socialism is the norm. Under the World State, men and women enjoy equality. The taboos of the Victorian Age – mostly those concerning sex – have disappeared. Hunger is unknown, poverty a memory. The cramped, filthy cities of nineteenth century England are replaced with gleaming spacious metropolises. Sickness has been practically eliminated and old age is no burden. Culture, education, sport are available to all. The practical, rational mind, no longer encumbered by the restraints of custom and tradition, has got to work and created a heaven on earth. Everyone is happy.

Yet, some of Wells remarks about his Utopia have a chilling ring to contemporary ears. He speaks of a future 'migratory population' and of a globalisation in which 'all local establishments, all definitions of place' were 'even now melting under our eyes'. 'Presently', Wells tells us, 'all the world will be awash with anonymous strange men'.[24] Our current obsession with security and identification is foreshadowed in Wells' belief that in the future World State 'there is every reason for assuming it possible that each human being could be given a distinct formula, a number or 'scientific name', under which he or she could be docketed'.[25] Wells' Utopia is a world, much like ours, under 'constant surveillance', through 'organised clairvoyance' producing an 'incessant stream of information' about its citizens.[26]

All of these developments are aimed at providing a 'universal security' and at eliminating what Wells saw as a 'cruel and wasteful wilderness of muddle'.[27] It is this 'muddle' that Wells and other social reformers in the early twentieth century saw as the greatest obstacle to improving human life, and their answer to it was to apply the scientific method, so successful in other realms, to life itself. We can say that our own world, informed as it is by science as the sole arbiter of truth, is a result of that belief. But not everyone was taken with Wells' vision, and for all its tidiness, reasonableness, and efficiency many readers today find it cold and unappealing. Even in Wells' day, intellects as reasonable as his own found fault with it. Aldous Huxley's *Brave New World* (1932), with George Orwell's *1984* (1949) the most well known dystopian work, is a satire on Wells' vision of a future World State.

An even earlier attack on the idea of a hyper-rational, scientific state appeared in 1907, the short story 'The Republic of the Southern Cross', by the Russian symbolist novelist and poet Valery Bryusov (1873–1924). Star City is a modern 'rational state' situated at the South Pole.[28] It is a perfect community. Everyone is happy, healthy, well-off, and in perfect harmony with each other. There is no envy or competition among its citizens. All live in exactly the same modern homes and wear the same modern dress. In Maslow's and Scheler's terms, their deficiency needs are all satisfied and their pleasure and vital values are all accommodated. Everything moves like clockwork and there is no muddle. But then a strange malady breaks out, a 'contradiction sickness'. People suddenly do exactly the opposite of what they intend. Madness quickly ensues, and the scientific city goes up in flames.

Most likely Bryusov did not know of Wells' *A Modern Utopia* but he would not have needed to in order to produce this chilling parable of the dangers of applying the gospel of science to life.

Bryusov's story is a long gloss on Dostoyevsky's dictum in *Notes From Underground* (1864), that if all of life could be arranged logically and rationally, a man would go insane on purpose, just to prove he was free. It seems that destruction, chaos, and death are preferable, even in the most perfect state, if through them we can feel we are free.

Strangers in a strange land

Working out a blueprint for an actual Eupsychia, then, may not be the best way to approach the question of becoming 'fully human'. Huxley made his own attempt at a Utopia in his last novel, *Island* (1962), a kind of reverse image of *Brave New World*. It is not particularly successful, either as a novel or as a social tract, although its advocacy of psychedelic drugs ('moksha medicine') as a means of self-development has endeared it to the counter-culture. Huxley's island Pala has everything a Eupsychia would want. The inhabitants meditate, enjoy guilt free love, are non-violent, non-competitive, are spiritual and free of neuroses. Yet there is ultimately something unsatisfying about it, as there is about all Utopias. And while going insane on purpose in order to show you are free is not a viable alternative, there is something in Dostoyevsky's rhetoric that suggests what is lacking.

The philosopher and psychologist William James (1842–1910), put his finger on it in his essay 'What Makes a Life Significant' (1899). James talks of a week he spent at the Assembly Grounds on Chautauqua Lake, a religious and adult education centre in upstate New York, established in 1874. James speaks of the 'sobriety and industry, intelligence and goodness, orderliness and ideality, prosperity and cheerfulness' that 'pervade the air.' He called it 'a serious and studious picnic on a gigantic scale ... equipped with means for satisfying all the necessary lower and most of the superfluous higher wants of man.' 'You have culture, you have kindness...you have equality, you have the best fruits of what mankind has fought and bled and striven for under the name of civilisation for centuries. You have, in short, a foretaste of what human society might be, were it all in the light, with no suffering and no dark corners.'

Yet there was a problem. 'Ouf! what a relief!' James found himself saying on leaving this idyllic setting. 'Now for something primordial and savage ... to set the balance straight again.' 'This order is too tame,' James wrote, 'this culture too second-rate, this goodness too uninspiring. This human drama without a villain or a pang ... this atrocious harmlessness of all things – I cannot abide with them. Let me take my chances again in the big outside worldly wilderness with all its sins and sufferings. There are the heights and depths, the precipices and the steep ideals, the gleams of the awful and the infinite; and there is more hope and help a thousand times than in this dead level and quintessence of every mediocrity.'[29]

What was wrong with this Utopia, James realised, was that it lacked 'the element that gives to the wicked outer world all its moral style'. What was missing was a sense of 'the everlasting battle of the powers of light with those of darkness.' It lacked the 'heroism' which, 'reduced to its bare chance' snatched 'victory from the jaws of death'. 'In this unspeakable Chautauqua there was no potentiality of death in sight anywhere, and no point of the compass visible from which danger might possibly appear. The ideal was so completely victorious already that no sign of any previous battle remained, the place just resting on its oars. But what our human emotions seem to require is the sight of the struggle going on.'[30]

In other words, James recognised, as did Rabbi Steinsaltz, that 'what makes a life significant' is the sense of challenge offered by the 'worst possible world in which there is yet hope'. Utopia is bland, but the struggle for it is invigorating. In the best of all possible worlds we would achieve nothing, an insight that those focussed on their deficiency needs and on the pleasurable and vital values find incomprehensible. The philosopher Johann Fichte (1762–1814) expressed this in his saying that: 'To be free is nothing. To become free is heavenly'. We require a challenge in order to get the best out of us, something James recognised in his essay 'The Moral Equivalent of War' (1906), and which Nietzsche expressed in his aphorism that: 'It is the good war that hallows every cause.' James and Nietzsche weren't militarists, nor did they mean to celebrate bloodshed and carnage. They recognised that struggle, resistance, effort, were keys to unlock our vital energies. Wells' Utopia, James' Chautauqua, Huxley's Pala, even Maslow's Eupsychia, may embody the values and aims we want to achieve, but without the actual fight for these, they seem unworthy of our regard.

It may be the case then, that in order to become 'fully human', in order to perform *tikkun* and take care of the universe, we may need to remain in this profoundly muddled world, a world in which we are not entirely at home, a world in which we are strangers suffering from a sometimes painful homesickness, a world that resists, at practically every turn, our efforts to transform it. But this doesn't mean that we should stop making our efforts. As Victor Frankl discovered (see Chapter 3), meaning can be drawn out of the worst possible conditions. If Frankl could find a reason to go on living and to help others find meaning in the midst of an Auschwitz and Dachau, we can easily find meaning and opportunities to perform *tikkun* in our everyday lives. We do not need the good society to be there, waiting for us, in order to become good, that is, responsible, people. And perhaps it is only by our being so that, at some point, the good society will arise.

Actualisation envy

I mention above that in his last years, Maslow was concerned with the question of why some people self-actualise, reach the level of being needs, while others don't. In other words, in Maslow's uncomfortable phrase, why are 'some people are more "human" than others', more 'fully human', that is? In an unpublished paper, Maslow voiced his concerns about what might happen when the distinction between actualisers and non-actualisers became apparent.[31] As Maslow believed that at bottom our capacity for self-actualisation was 'instinctoid', then the difference between actualisers and non-actualisers was 'natural' and not a question of society or the environment, a view of things that runs against our contemporary thinking.[32] Maslow was concerned that 'when there is no longer social injustice to serve as an excuse for one's own biological inadequacies ... there might well be a great increase [of] malicious envy of those who are more successful in their achievements.' (Scheler voiced a similar concern regarding his concept of *ressentiment*.) Thinking of ways to 'protect the biologically gifted from the almost inevitable malice of the biologically non-gifted', Maslow suggested something like a new 'priestly class to which is given less monetary reward and fewer privileges and luxuries than the average members of the overall population,' given that self-actualisers are less interested in material rewards than in the 'metagratifications' or 'intrinsic values' of 'advancing beauty, excellence, justice, or truth'.

In Maslow's fantasy, non-actualisers, then, would enjoy the fulfilment of their deficiency needs without limit, their pleasure and vital values would be met completely, while actualisers would be left to pursue their being needs and spiritual-cultural values unhindered. There is no sense in which someone's 'being needs' would be deprived. There is no group that 'gets to do this' while others don't. No one decides who should actualise and who should not. No one is kept from actualising those needs should they appear. The idea of non-actualisers demanding their 'actualisation rights' would be ludicrous, as that would mean, say, that someone with no interest in philosophy would demand the right to spend his time reading Plato. What Maslow wants to protect in his fantasy, are the rights of those who do want to read Plato, their right to do so, without being subject to the possible scorn of those that don't.[33]

Maslow uses difficult language for us, and even speaks of a 'biological élite', a term that smacks of dark connotations, of eugenics and breeding. But he is really only speaking of recognising something that is happening anyway, in 'nature', and of possible ways of avoiding social friction. The term élite, however, disturbs us. But Maslow's élite are not interested in profiting from their distinctions. As mentioned earlier, self-actualisers, the 'fully human', are not interested in expressing their actualisation, their 'full humanness', through dominating others. They are too busy exploring their being-needs, their spiritual-cultural values, and in any case, any gesture of domination would, by definition, be one of deficiency, of being 'less human'. Maslow had no interest in establishing an 'us and them' scenario, and if Wells is right, and a new kind of person has appeared in the last few centuries, driven by 'creative' needs, then perhaps there are more self-actualisers than Maslow thought.

Wells himself, though, recognised the need for a kind of élite to manage things in his modern Utopia. He called them the Samurai. Scheler, who devoted a great deal of thought to the possibility of a society run along spiritual-cultural lines, also believed that it would naturally produce a small group of leaders, who would set the values and guidelines for the rest. Colin Wilson writes of the 'outsiders', men and women driven by deep spiritual needs who, ignored for the most part by society – if not actually opposed by it – nevertheless produce values and creative energies that push society to evolve. In one of his last books, *Two Sources of Morality and Religion* (1935), Bergson spoke of a 'creative minority' whose breakthroughs into new areas of insight and inspiration guide society, an idea adopted by the historian Arnold Toynbee in his monumental *A Study of History* (1934–1961).

The creative minority

Our egalitarian sensibilities find these notions dubious, but that is no real argument against them. And in any case, our egalitarian world is full of élites, of celebrities, sports figures, millionaires, the people whose lives we envy and want to enjoy and who we read about in tabloids and gossip magazines. And again, we have to recognise that the creative individuals Maslow, Scheler and the others speak of are not interested in other people, and certainly not in dominating them. This is in fact something that sets them apart. One of the characteristics of self-actualisers, Maslow discovered, is that they prefer solitude to gregariousness, and most times prefer their own company to that of others. Not out of a dislike for people, a haughty misanthropy, the kind of anti-human sentiment that informs Lovecraft's work. But because their own interests, their being needs, can only be pursued on their own, not in a crowd. They feel more at home in their inner world than in the social one around them. Their dominance is expressed in their creativity, not in lording it over others, and those who need others in order to express their dominance – pop stars, celebrities, politicians, gurus – are, in fact, altogether less dominant, because they collapse without that explicit validation. In *The Nature of Sympathy* (1922), his work on what binds us together in a community, Scheler wrote that: 'Gregariousness in animals represents an advance toward the human level, whereas man becomes more of an animal by associating himself with the crowd, and more of a man by cultivating his spiritual independence'.[34] The more human we are, the less we need other people. There is no such thing as a self-actualising dictator, as by definition a dictator requires other people, lots of them. This is in line with passing beyond the sphere of vital or life values and entering the spiritual-cultural one. It is also in line with reaching Maslow's being needs.

I do believe there is a creative minority at work today, trying to become 'fully human', looking for ways to repair the universe, and trying to do the good that they know. But they are a silent minority. Berdyaev, pre-echoing Kazantzakis, says that 'the way to all creativity lies through readiness to sacrifice'.[35] 'The move to creativeness in the social order is possible,' he wrote' only by sacrificing security and guaranteed well-being', that is, the pleasure and vital values, the anxiety over deficiency needs. 'This readiness to sacrifice', Berdyaev said, 'has nothing in common with anarchy, with chaos' – it is not the destruction of the social order – but 'is always cosmic in character'. Such a creative society, Berdyaev believed, is 'subterranean'.[36] Its members work below the surface and do not draw a lot of attention to themselves. Years ago Wells spoke of what he called the Open Conspiracy. It is something along these lines that I have in mind.

I agree with Berdyaev and I think Maslow's ideas bear him out. I think that society as a whole has reached the level of Maslow's self-esteem needs, at least in the affluent west, where most people are well-fed and housed. My evidence for this is the global popularity of social networks. These online communities make manifest Andy Warhol's oft-quoted remark that 'in the future everyone will be world famous for fifteen minutes'. These, and other developments, such as 'reality TV', in which 'real' people are the celebrities, suggest to me that our self-esteem need is being satisfied at a planetary level. We post photographs, remarks, quotations, and videos online that, in the past, we could share only with our immediate friends. Now they reach around the world almost instantaneously. A video of someone talking to his dog can go 'viral' and command world wide attention. People just like us are the focus of millions of other people's attention on programs such as *Big Brother* (again, Wells' 'constant surveillance', through 'organised clairvoyance' producing an 'incessant stream of information' about his World State's citizens). We are stars. We are somebody. We matter. It is no secret that many people spend much time and effort coming up with gimmicks that will get the online world's attention. We can have hundreds, even thousands, of close friends, with whom we can share practically everything about our lives, and so receive instant validation for ourselves. I think that with these shifts in our social world, our self-esteem needs and the vital, life values, have reached a kind of apex. We are all somebody now, and everybody knows it.

My silent minority, the members of Berdyaev's subterranean creative society, have moved past this level and have entered that of the being needs and the spiritual-cultural values. They no longer need their self-esteem topped up repeatedly, nor to feel connected to others constantly. They do not tell everyone about the good they did recently, or about what bit of the cosmos they are going to repair. They do not occupy Wall Street or anywhere else, but their minds. They might not know of each other, but that doesn't matter. There's work to do, and they do it.

5. Beyond Nature

If the idea of 'taking care of the cosmos' or 'repairing the universe' elicits a response in people at all, their first thought will more than likely be of nature. What is in more need of care or repair then the planet that we have ripped, torn, and dug into for millennia? What else can we save but it? The idea that nature is at risk, has been abused, and needs our help has been a part of modern consciousness for at least half a century now. We can mark the publication of Rachel Carson's *Silent Spring* (1962) as the beginning of a wide, popular awareness of our ecological and environmental responsibility, a responsibility that more and more people are willing to embrace. That can only be for the good, and only a fool could fail to recognise the immense debt we owe our long suffering Mother Earth. Yet, if our sensitivity to questions about society, egalitarianism, and élites makes asking certain questions and voicing certain concerns about the 'fully human' uncomfortable and risky, our sensitivity to nature makes expressing certain ideas and thoughts about our relation to the natural world even more difficult.

We are often self-righteous and defensive when it comes to questions about any possible 'natural' differences between people, differences, that is, that cannot be accounted for in terms of environment, society, class, privilege, or other controllable artificial factors. But paradoxically we are even more sensitive to any comments that question the primacy and importance of nature and 'the natural'. It is extremely difficult, in our anti-human times, to voice any reservations about our celebration of nature and our need to 'get back' to it, without them being taken the wrong way. If, as I mention in Chapter 3, sex in our time is an unquestioned good – perhaps the sole unquestioned good – nature has in many ways been adopted as an almost universally unquestioned noble cause, as well as a cure for most, if not all of our ills.

We have lost touch with nature. We have savaged it. We have fallen out of harmony with it. We have ignored it. We have acted selfishly toward it. We have despoiled it, we have raped it, and we have, in many instances, even destroyed it. We want more and more nature back in our lives, and human interference in such things as food (GM) is seen as almost axiomatically aberrant. Nature, for many of us, is seen as all good and the source of all good, while human interference with the natural world is seen as almost always bad. Documentaries about nature, television shows about animals 'in the wild', communicate how vital our link to the natural world is, and how we need to find our place within the harmonious nature of things, and how we are, after all, only animals, and like them, just a part of nature. Nature is our friend, and we should do more and more to be at home with it. At this point I am not asking whether this brief rundown of our ideas about nature is true or not, merely characterising what strikes me as a dominant sensibility of our time.

How much this celebration and valorisation of nature is born of an accurate understanding of it and how much is rooted in our current antipathy to the human is debatable. We trust nature where we do not trust the human, and in many, perhaps most instances, that trust is justified. Human beings make mistakes while nature does not. Poisonous insects and hurricanes are not mistakes; they are merely parts of nature that do not agree with us. They are natural, as are tsunamis and earthquakes. I mention these somewhat clichéd examples of a less nice and harmonious nature not in order to argue against nature – that would be as pointless as arguing against clouds – but in order to show that our wholesale embrace of the natural is actually rather selective. 'Nature' is not a monolithic, homogenous something we encounter when we leave the city and head to the country, but is in many ways a product of our interpretations – as is, in fact, practically everything.

By this, of course, I do not mean that trees, hills, oceans, stars, algae, nightingales, and other forms of nature do not exist in themselves. But what these natural things mean to us depends on our relationship to them, how we understand them, our attitude toward them, and the kind of consciousness we bring to them. As Iain McGilchrist says, and as I've already quoted, 'the kind of attention we bring to bear on the world changes the nature of the world we attend to'.[1] (Or, as Blake said with characteristic clarity: 'The Sun's light, When he unfolds it/Depends on the Organ that beholds it.') With McGilchrist I believe that 'there *is* something that exists apart from ourselves' – a statement that in one sense seems unassailably obvious, yet, for the reflective mind is not as easy to prove as it may seem – 'but that we play a vital part in bringing it into being'.[2] In this sense there is more than one nature. We think it 'unnatural' if we modify grains genetically, but we would think it 'only natural' to want to come up with a means of modifying tidal waves or mud slides so that they would be less dangerous. In one case, pure, untouched, pristine nature is a good, whereas in the other it is not. This is not to say that nature is bad because of earthquakes and tsunamis – nature is neither good nor bad, only human beings are – merely that much of our appreciation of nature and our often unthinking championing of it may be born of a kind of sentimental idea of it, a kitsch version in which those parts that are pleasant and agreeable are emphasised and the less salubrious portions ignored. In many ways the nature that we enjoy and look to as a panacea of our ills is a nature that fits in with our human needs and purposes. Although many of us want to get back to nature, the nature we get back to is unavoidably human.

The discovery of nature

We can even say that the nature we want to get back to is an almost purely human creation. It is a relatively recent invention, a product, we can say, of the evolution of consciousness. For most of human existence, nature was an unwieldy, difficult and uncertain taskmaster, demanding constant effort with little return. Nature then was tree stumps, boulders, barren soil, locusts, and floods, much more then it was beautiful meadows and entrancing forests. As we saw in the last chapter, H.G. Wells pointed out that for most of history, human beings have been 'up against it', and what they have been up against is nature. And although nature is a theme in the poetry of earlier times – in Virgil's *Georgics* (29 BC), for example, whose theme is agriculture – there was not the kind of love of nature and disinterested contemplation of it that we associate with our own conceptions of it. This view of nature began in earnest little more than two centuries ago, although its roots in the evolution of consciousness reach back some centuries earlier. We can say that the origins of our modern appreciation of nature go back to 26 April 1336, when the Italian poet Francesco Petrarca (1304–1374), better known as Petrarch, made his famous ascent of Mount Ventoux in France. This event has gone down in history as the first time someone climbed a mountain solely to see the view. Clearly people had scaled heights before, but Petrarch claimed he was the first to do so solely out of curiosity, for what we might call aesthetic reasons. He recounted his excursion in one of the letters making up his *Epistolae familiares* (1350), and I write about it at length elsewhere.[3]

Many thinkers have commented on the significance of Petrarch's ascent. For the nineteenth century German historian of culture Jacob Burckhardt (1818–1897), it was in Petrarch that nature's 'significance ... for a receptive spirit is fully and clearly displayed', and because of this he was 'one of the first truly modern men'. Burckhardt's choice of words in describing the importance of Petrarch's ascent links it to the *Sehnsucht* that would trouble the poets of a later time. Burckhardt saw in Petrarch 'an indefinable longing for a distant panorama' that 'grew stronger and stronger in him' and was the central motivation for his excursion.[4] Ernst Cassirer saw in Petrarch's ascent of Mount Ventoux 'testimony to [the] decisive change in the concept of nature that began in the thirteen century' and which led to nature becoming a 'a new means of expression' for human consciousness, as well as to a 'desire to immediately contemplate nature'.[5] More recently, the late archetypal psychologist James Hillman (1926–2011) saw in Petrarch 'the complexity and mystery of the man-psyche relationship', in which we may 'turn outward to the mountains and plains and seas or inward to images corresponding with these'. For Hillman, Petrarch is both the 'discoverer' of nature and of the reality of the inner world as well.[6]

Whatever the philosophical significance of Petrarch's ascent, it presaged our contemporary delight in mountain landscapes, in hiking, and in trekking out to inhospitable and almost inaccessible locations.

It took some time, however, for the practice of seeking out vistas and scenic views to catch on, and even as recently as the late eighteenth century, nature as we understand it, as an object of aesthetic contemplation, did not really exist. In 1773, Dr Samuel Johnson (1709–1784), made his celebrated journey to Scotland, chronicled in *A Journey to the Western Isles of Scotland* (1775). In his account, Johnson often complains about the numerous lakes and mountains his carriage had to circumvent. He said they made the journey tedious and long. Johnson was adamantly urban – famously he said that 'when a man is tired of London, he is tired of life' – and he admits that his reason for taking his trip was to see 'the wild', the same reason we give for our weekend camping adventures. But for Johnson it was a one-off and he returned chastened and somewhat disappointed. We belonged in civilisation, Johnson believed, not the wilderness, and the proper study of mankind, as Johnson's contemporary Alexander Pope declared, was man, not the wastes. A beautiful garden in that time was a well sculpted topiary.

The Romantics

Today people spend a great deal of time, money, and effort going out of their way to get to precisely the kinds of 'nature spots' that a few centuries ago Dr Johnson and practically everyone else would have assiduously avoided. When Petrarch informed an old shepherd about his intention to climb Mount Ventoux, the shepherd warned him not to, saying the idea was mad, even demonic, and would only bring disaster. 'Nature' as we know it, then, has a history. If we want to give a date for its debut, we can say that the nature that we know and love and which we make great efforts to embrace first arrived in 1798 with the publication of the *Lyrical Ballads* by the poets William Wordsworth and Samuel Taylor Coleridge. Wordsworth and Coleridge's seminal collection introduced the Romantic sensibility to English speaking readers, and it is no exaggeration to say, as the literary philosopher Owen Barfield (1898–1997) has, that the holiday industry, which offers trips to mountains, forests, deserts, and other uncivilised places, owes a great deal to the shift in human consciousness exemplified in their work. If you ever doubted the ability of the imagination to change the world, take a look at Thomas Cook and the *Lonely Planet Guide*, then think of Wordsworth's 'one impulse from a vernal wood'.[7]

Of course earlier voices celebrated the virtues of nature. In his *Discourse on Inequality* (1754), the Swiss novelist and social philosopher Jean Jacques Rousseau (1712–1778) argued that civilisation was the cause of man's woes; in his 'natural' state, man was good, Rousseau believed, and he became evil only through the influence of decadent society. Rousseau's ideas spread the myth of the 'noble savage' – although Rousseau himself never used the term – which is still popular today, as can be seen in our romantic conceptions about 'indigenous peoples' untouched by modernity's evils. (These tell us, I think, more about our own guilty conscience than the reality of these people's lives.) And Goethe more or less started the Romantic movement with his novel of unrequited love and suicide, *The Sorrows of Young Werther* (1774), whose troubled hero is given to sudden fits of ecstasy in the face of the nature that Dr Johnson found tedious.

In 1796, two years before Wordsworth and Coleridge's *Lyrical Ballads*, Mary Wollstonecraft (1759–1797), author of *A Vindication of the Rights of Woman* (1792), published a travel book, *Letters Written During a Short Residence in Sweden, Norway, and Denmark*, in which she anticipates the kind of contemplative delight in nature that Wordsworth and Coleridge would make widely popular. Contemplating the rugged scenery of the northern lands, she speaks of 'that spontaneous pleasure which gives credibility to our expectations of happiness' and saw that 'the sublime often gave place imperceptibly to the beautiful, dilating the emotions which were painfully concentrated'.[8] The 'sublime' as contrasted with the beautiful – the terror and awe we feel in the face of something overwhelming – was first contemplated by Longinus in *On the Sublime* (1 AD), and in Wollstonecraft's day had become a popular theme through Edmund Burke's influential essay *A Philosophical Enquiry into the Origins of our Ideas About the Sublime* (1756). Where classical beauty was clear and distinct, and its effects generally agreeable, the sublime partook of the mysterious, the alien, and the weird, and it could terrify; hence its attraction to the Romantics, and to moderns like the poet Rilke for whom 'beauty is nothing but the beginning of a terror, which we are still just able to endure'.[9] For Burke and his readers, the 'sublime' was perceived in mountains, thunderstorms, and other manifestations of wild nature. On approaching the Swedish coast, Wollstonecraft saw 'that the sunbeams that played on the ocean, scarcely ruffled by the lightest breeze, contrasted with the huge dark rocks, that looked like the rude materials of creation, forming the barrier of unwrought space'. 'This view of the wild coast' – Scandinavia was considered more or less *terra incognita* then – 'afforded me continual subject for meditation' a remark that Dr Johnson may well have snorted at.[10]

New outer and inner worlds

As you might expect, exactly why human consciousness 'discovered' nature as a source of contemplative delight in the last decade of the eighteenth century, is a complex question. And it should be clear that when I speak of 'nature' what I have in mind is its scenic, picturesque character, not the nature 'red in tooth and claw' (Tennyson) that Aldous Huxley took Wordsworth to task over in his essay 'Wordsworth in the Tropics' (1928). Here Huxley argues that Wordsworth's unquestioned love of nature is possible only where nature herself has been tamed, something I suggest at the beginning of this chapter. This itself tells us that part of the answer to the question why we discovered nature roughly two centuries ago, must surely be the fact that, as H.G. Wells remarked, by then humanity had more or less won its difficult battle against 'brute nature' and could now step back from it and regard it at leisure. We can say that until then we regarded nature solely with our 'survival consciousness', as something we had to deal with or avoid. We were, as Wells says, 'up against it'. With some measure of victory, we could draw back, and brute nature, then, was no longer 'in our face'. We had achieved some distance from it.

Not everyone, of course. The farmers who regarded Wordsworth with curiosity as he gazed meditatively on the fields did not share in his new liberation from ploughs and furrows. But the change had started. Colin Wilson argues that what had happened with the Romantics is that Western man had discovered the power of the imagination, that with them consciousness was no longer 'stuck' in the present moment, which, for most people, meant being smack up against brute nature. We had found ourselves on the shores of a new, inner world, just as the explorers of a century or two earlier had found themselves on the shores of new geographical lands. Of course, as with nature itself, people had been aware of the imagination ever since early hunters sat around the campfire and recalled the excitement of the day's kill, or listened to old warriors recount battles from long ago. But what seems to have happened is that in the late eighteenth century, this ability to enter our own minds was somehow increased, as if after centuries of getting by on only slightly fermented grape juice, we were suddenly introduced to 80 proof alcohol. It was a powerful new drug, and we took to it.

Exactly why this happened is unclear, and the attempts to explain it in economic, sociological, political, and other rational, ultimately material terms seem, to me, to fall short. They leave out the possibility that life itself, driven by its own inherent need to grow beyond itself – Bergson's 'creative evolution' – pushed consciousness into these new spheres, in the same way that my inherent curiosity and need to develop pushes my own consciousness into new interests and desires. Whatever the reason, we can say that in many ways, the discovery of nature – of the outer, physical world as an object of contemplation – came about because of the new inner territories that had opened up.[11] What happened with the Romantics is the same thing that the poet Rilke experienced a century later. 'I am learning to see,' Rilke wrote in his existential novel *The Notebooks of Malte Laurids Briggs* (1910), about his time in Paris in the early 1900s. 'I don't know why, but everything enters me more deeply and doesn't stop where it once used to. I have an interior that I never knew of.'[12]

A new perspective

The philosopher Jean Gebser also believed that by the time of the Romantics, human consciousness was undergoing a decisive change, and that this affected the way we saw the world around us, had produced, as it were, a new kind of nature. For Gebser, the shift had begun with Petrarch's ascent of Mount Ventoux. With this, according to Gebser, human consciousness had lifted itself out of the flat, two-dimensional world of the Middle Ages and had discovered 'space' for the first time. In his account of the ascent, Petrarch speaks of seeing 'the clouds lay beneath my feet', a sight that only the ancient gods of Olympus had hitherto enjoyed. From his new vantage point Petrarch could see the mountains of Lyon, the Mediterranean, Marseilles, the Rhone, and in his account he is not a little terrified – in the grip of the sublime – by the spectacle, which in many ways he believed is an act against God. For Gebser, Petrarch's experience was an 'epochal event' which signified nothing less than 'the discovery of landscape; the first dawning of an awareness of space that resulted in a fundamental alteration of European man's attitude in and toward the world.'[13] From Gebser's point of view, then, the 'space age' began in 1336, and not in 1957 with the Russian satellite Sputnik. In fact, according to Gebser, the whole metaphor of having a 'point of view' could not have arisen until this change in consciousness had taken place. What began with Petrarch's ascent, for Gebser, was the age of what he called 'perspectival consciousness', the perception and representation of the world from a unique human vantage point. If we contrast the tapestries characteristic of the medieval period with the perspective paintings that came to prominence in the Renaissance, we can, I think, see what Gebser means.

In the medieval world figures seem 'embedded' in the landscape; there is no sense of distance or depth in medieval tapestries, just as there is very little of it in Egyptian wall paintings. The kind of space that O.V. de Lubicz Milosz despaired of was not perceivable to medieval man, at least according to Gebser. People of the Middle Ages, he believed, felt they were part of a kind of 'fabric of being', and that there was an immediate, almost tactile continuity between themselves and the world around them. In *Saving the Appearances* (1957), a work devoted to the idea of an 'evolution of consciousness', Owen Barfield gives us a glimpse of what the experience of being in the world may have been like for medieval man. Writing as if he was a 'medieval man in the street', Barfield suggests that when he looks at the sky, he does not 'see it as empty space'. 'If it is daytime, we see the air filled with light proceeding from a living sun, rather as our own flesh is filled with blood proceeding from a living heart. If it is night-time, we do not merely see a plain, homogenous vault pricked with separate points of light, but a regional qualitative sky, from which first all of the different sections of the great zodiacal belt, and secondly the planets and the moon ... are raying down their complex influences on the earth ...We take it for granted that those invisible spheres [of the planets] are giving forth inaudible music ['the music of the spheres']...We know very well that growing things are specially beholden to the moon, that gold and silver draw their virtue from the sun and moon respectively, copper from Venus, iron from Mars, lead from Saturn, and that our own health and temperament are joined by invisible threads to these heavenly bodies we are looking at.'[14] It is in this sense that Barfield suggests that medieval or 'pre-perspectival man' did not feel that he was *within* space. He wore his world around him like a garment, rather than moved about it as on a stage, or, perhaps more apt for ourselves, rattled around in it as objects in a box. According to Barfield, nature, the world, was something medieval man, and his ancestors, *participated* in, not something detached from them and held, as it were, at arm's length.

With Petrarch's ascent of Mount Ventoux, Gebser believes, all this changed. Human consciousness suddenly became aware of distance, of depth, and of what we all take for granted as 'empty' space. The closely woven intertwining between things began to unravel. Where before a felt relationship held things in place, and allowed for a flow of being, now vast vistas and chasms had opened up, and in the new 'empty' space, human consciousness felt its moorings loosen. In much the same way as Copernicus' heliocentric solar system 'sent the Earth rolling toward X', Petrarch's daring ascent had unhinged man from his 'place' and set him free to roam. This change in the perceived world was paralleled by changes in how man perceived his place in the wider, spiritual cosmos; hence the celebration of man and human virtues – of the past and its knowledge – that characterises the period we know as the Renaissance. Whereas medieval man felt himself to be a sinner, corrupt, unworthy, in constant fear of falling into the Devil's hands, and one creature among many on God's green earth – although one with a particular destiny and position – the new dispensation championed man's creativeness, his daring, his freedom and almost godlike potential. Nothing expresses this shift in man's sense of himself as forcefully as the title of one of the most important works of the Renaissance, Pico della Mirandola's *Oration on the Dignity of Man* (1486). Human dignity was not something commonly accepted in the previous tapestry of life, but for the fiery, independent personalities of the Renaissance, it was an unquestioned good.

A change of nature

'Nature', then, is not something independent of human beings, something 'out there' that we passively reflect in our consciousness, as a mirror passively reflects the objects in front of it, and which would be there, just as it is, if we weren't observing it. This is the attitude toward the world that Husserl called 'the natural standpoint': the unreflective acceptance of 'the world' as given, without an awareness of the mind's own contribution to it. Husserl called the mind's contribution its 'intentionality', its selective, purposive, active character. As Colin Wilson has expressed it, rather than simply mirror what is there, the mind, according to Husserl, 'reaches out' and 'grabs' it. No mirror is selective, nor does it reach out and grab the objects it reflects. This contribution takes place at levels of the mind we are not ordinarily conscious of, yet our attitudes and expectations inform it. We are not immediately aware of this and the insight, once grasped, is difficult to hold on to, so strong is the pull of seeing consciousness as merely a mirror, or a camera, an analogy rooted in our passive, Cartesian habits of thought. For the moment let us accept this and try to hold on to it firmly: the 'nature' we encounter depends on the consciousness we direct at it. Or, as Blake put it, 'in your Bosom you bear your Heaven and Earth & all you behold; tho' it appears Without, it is Within, in your Imagination, of which this World of mortality is but a Shadow'.[15]

It may seem a loss to have shifted from a world we wore as a garment to one we move about in as on a stage, from a continuous, interwoven world, to an empty, disconnected one. But while our earlier 'participated world' was more connected, it was also one in which we were less free. A figure on a tapestry, no matter how beautiful, is fixed. And however much we may feel that the cosier, more intimate world of the Middle Ages is desirable, it had its drawbacks. Barfield's picture of how medieval consciousness may have perceived its world may suggest a state of mind preferable to our own, isolated, detached awareness. But it was a world in which consciousness was embedded. And the same was true of earlier times. Our contemporary fascination with 'primitive' peoples, and with the people of ancient civilisations, who, we believe, were closer to nature than we are, is understandable, but it ignores the fact that most likely the people of earlier times were not able to *think* about these things in the way that we can, did not, in fact, feel themselves to be an 'I' in the way we do. They were not able to step back and regard the world as something 'other', at least not in the way that we can. In this sense they existed in the kind of ouroboric embrace we enjoyed as children, but remained in it all their lives. In fact, the kind of intimacy people of an earlier time felt with the world around them is most likely something we feel in our childhood, when everything, even inanimate objects, seems to be alive and responsive to us. Yet none of us, for all our nostalgia, think it a good idea to remain children all our life – our current overestimation of youth and youthfulness notwithstanding. As Barfield says, medieval man – and, by inference, humankind from earlier periods – did not reflect on his experience. He took it for granted. He lived in a much more living world, but it was one which we never questioned, and which held him firmly in place.

Leaving home

Petrarch's ascent of Mount Ventoux, according to Gebser, freed us from this warm, but ultimately limiting embrace. But it did so at a price. One of the results of Petrarch's daring adventure, Gebser argues, was the kind of universe that frightened Pascal with its infinite silences and vast empty spaces. Gebser's reading of the importance of Petrarch's ascent is in the context of his philosophy of the 'structures of consciousness', decisive shifts in the character of human consciousness that have taken place throughout history. Elsewhere I have written about Gebser's structures of consciousness in the context of other ideas about the evolution of consciousness, including Barfield's, and the reader can find a fuller exposition of Gebser's and Barfield's ideas there.[16] Here I can only say that for Barfield, Gebser, and other thinkers I discuss, this separation from our earlier, embedded relation to the world, into our more isolated, disconnected one, was a necessary step in our evolution, the equivalent, in terms of our consciousness, of a child realising it is a separate, independent being, no longer completely protected from life's dangers by its parents. With this recognition comes a sense of freedom and adventure, but it also brings with it an often overwhelming fear.

In terms of the history of western consciousness, we can see the period of the Renaissance as the expression of the sense of adventure, of the opening up of new vistas and broad, uncharted depths, and the confidence to explore them. But it is precisely those uncharted regions that, as the modern period progressed, fed a growing hint of fear, or at least of disquiet, settling, in our time, into a profound sense of cosmic inconsequence. For Gebser, Petrarch's ascent of Mount Ventoux marked what he called the 'deficient mode of the mental-rational consciousness structure'. The 'mental-rational structure', which Gebser believes began to emerge at around 1225 BC – Gebser's dates are always estimates, but he gathers an impressive amount of evidence to support them – is one in which consciousness begins to experience the first signs of 'directed' or 'discursive' thought.[17] In the previous structure, which Gebser calls the 'mythic', thought was more involved with images; it was more along the lines of what Rudolf Steiner called 'picture-thinking' and which he links to what he calls 'old moon consciousness'.[18] (In *A Secret History of Consciousness* I write at length about the clear parallels between Gebser's 'structures' and Steiner's 'epochs'.[19]) Now consciousness begins to feel an independence from the world around it, and the individual begins to stand on his own, no longer completely supported by the kind of group consciousness characteristic of earlier ages.

A dangerous freedom

The clearest sign of this new, mental-rational structure for Gebser is the rise of Greek philosophy, and in many ways his interpretation of this is similar to Heidegger's, who sees in the rise of Platonic metaphysics the roots of what he calls 'the age of the world as view', the age, that is, of the world as something to be seen and analysed. Rational, logical, discursive thought, the search for answers, for explanations, the roots of science, begin to take hold at this point, and the mental independence necessary for the evolution of consciousness is firmly in place. With the rise of the mental-rational structure, we emerged as creatures who, for the first time, felt on our own in the world. This, for Gebser, is both a blessing and a curse, because it at once provides us with the necessary freedom to grow and develop, but also makes possible a complete separation from what he calls 'origin', the timeless, non-spatial, immaterial, spiritual source of the different structures of consciousness (we can say that in essentials Gebser's 'origin' is equivalent to the Kabbalah's *Ein-Sof*) The rise of the mental-rational structure, Gebser says, is the beginning of our estrangement from the world, a loss of connection with it which by now we take for granted. What happened when Petrarch made his mad ascent of Mount Ventoux, is that the mental-rational structure entered what Gesber calls its 'deficient mode'. This is when the characteristics of a particular structure of consciousness, which were previously an asset, become a liability. By the time of Petrarch's ascent, the detachment from the world that enabled the rise of philosophy and science – of *thought* itself – was moving into the kind of alienation from it common to our own time.

As the world was seen more and more as something distinctly apart from ourselves, something we could hold at arm's length and perceive as an object, its living, vital, participatory character dimmed. We can say that as our own inner world grew – our own sense of being an independent, conscious self – the world's 'interior', its soul, faded. By the time that Pascal expressed his fear at 'the eternal silences of the infinite spaces' the participatory, living cosmos was on its way to being completely eclipsed by the new, anti-animistic vision ('world as view') of the world as a great machine, operating under mathematical law – a vision we still embrace today. In *The Quest for Hermes Trimegistus* I mark the triumph of this 'modern' view in the work of the Catholic monk and philosopher Marin Mersenne (1588–1648), an older contemporary of Pascal and colleague of René Descartes, commonly seen as one of the founders of our scientific world view. Mersenne is important because he initiated an unequivocal attack on the kind of world medieval man had experienced, and which was still being contemplated in his day by followers of the Hermetic teachings, that is, a living, participatory world.[20] This ability to stand back and regard the world as something 'other' led to an increase in man's confidence in his own powers and abilities. Yet while this initially led to a powerful resurgence in Hermetic and spiritual philosophy – the recovery of the *Corpus Hermeticum* and Plato's works were responsible for this and were more or less what the Renaissance was about – it slowly transformed into the atrophying of his new found power to view the world as an object (Gebser's 'deficient mode of the mental-rational structure'). This then quickly shifted into having power *over* the world, expressed in Francis Bacon's dictum that 'knowledge is power' (although Bacon himself apparently never phrased it quite in that way), and which we can see as the beginning of the modern dominance of the left-brain approach to reality. From a garment we wore and a world we participated in, nature and all its works became something we could control through the power of mathematics.

A ruler in chains

While this development brought undoubted benefits which we still enjoy – it would be tedious and unnecessary to list all the advantages we owe to the rise of science and technology – it resulted in a world and a nature that was lifeless, mechanical, and shot through with a cold, implacable necessity. It also eventually led to the purposeless universe of Steven Weinberg and the pointless drift of John Gray. Nature and everything in it followed irrevocable routines, whether we wanted them to or not. It was a great clock, ticking away, all its cogs and wheels fitting perfectly into each other. God may have created it to begin with – if nothing else it was politic to state this – but once he did, it ran without his interference, much as Stephen Hawking's universe created itself without help from anything. Physical laws and mathematical necessity took care of everything, and by the early 1800s, the French astronomer Pierre-Simon Laplace argued that if we could know the exact location and momentum of every particle in the universe we could predict with absolute certainty the future movements of both the galaxies and the atoms of which they were composed; this was, in fact, an early 'theory of everything'. The present grows from the past as unavoidably as the future does from the present; the inexorable chain of cause and effect ensured this. The notion of free will and the idea that human beings were, by divine dispensation, exempt from such inescapable determinism, looked pretty threadbare. Man had freed himself from the superstitions of the past, had mastered nature, and enthroned himself as the new ruler of the universe, but he was a ruler in chains, subject to the very iron laws that had enabled him to establish control.

It was against this new, clockwork nature that minds unhappy with the scientistic view rebelled. I say 'scientistic', not scientific, because almost from the beginning this new way of seeing the world hardened into a dogma, into a new religion, 'scientism', which professed belief in this new, mechanical way of perceiving the world as the only valid one and in its results as being the sole criteria for truth. Truth was what you could perceive with the senses, hold in your hands, and weigh and measure. What you could not, could not be proved. Needless to say, most of the truths that make life worth living could not meet this requirement – how much does love weigh? – and so, as we saw earlier, they were relegated to the status of un-truths.

Against nature

One of the most vocal opponents of the new mechanistic vision of nature was William Blake. In Great Britain today Blake is seen as a national poet. His poem that prefaces his prophetic book *Milton*, beginning with 'And did those feet in ancient times', as set to music by Sir Hubert Parry, is known as the patriotic anthem 'Jerusalem'. But in his own time Blake was quite literally considered mad, and his poems were read by just a few followers. Perhaps this was not unreasonable, as Blake was a follower of Swedenborg, and Swedenborg's own sanity was questioned; at one point an attempt was made to have Swedenborg declared insane and intered in an asylum.[21] Blake's madness was an expression of his visionary faculty, his ability to see a world in a grain of sand or heaven in a wild flower. In *A Vision of the Last Judgement* (1808) Blake writes, 'it will be Questioned "When the sun rises, do you not see a round disc of fire somewhat like a guinea?" His answer is: 'O no, no I see an Innumerable company of the Heavenly host crying Holy Holy Holy is the Lord God Almighty' – his vision of the 'spiritual sun'.[22] Blake knew that the mind, or, in his terminology, the Imagination, was responsible for the sun, the stars, the whole 'Outward Creation', and that, as quoted above, although it appears without, it is within. 'Mental Things are alone Real' – I retain Blake's idiosyncratic spelling and punctuation – 'what is Calld Corporeal Nobody Knows of its Dwelling Place'. 'The Nature of Visionary Fancy or Imagination is very little Known & the Eternal nature & permanence of its ever Existent Images [Gebser's 'Ever Present Origin'?] is considered as less permanent than the things of Vegetative and Generative Nature yet the Oak dies as well as the Lettuce but Its Eternal Image & Individuality never dies.' 'This world of Imagination is the World of Eternity it is the Divine bosom into which we shall all go after the death of the Vegetated body'. 'There Exist in that Eternal World the Permanent Realities of Every Thing which we see reflected in this Vegetable Glass of Nature.'[23]

For Blake, as for Swedenborg, Plato, and innumerable mystics, the outer world is a reflection or symbol of an inner one. For them the real world is this higher, spiritual, interior reality, and the lower, physical universe is merely a sign of it, or a kind of language, which we can learn to read. As Kathleen Raine puts it, speaking of Blake and Swedenborg, 'It is not mind but nature which is the mirror, in which objects have only an apparent existence.'[24] As Blake himself says, 'Where man is not Nature is barren' ('Proverbs of Hell').Yet neither Blake's nor Swedenborg's vision is anti-nature in some simplistic sense. Blake tells us that 'a robin redbreast in a cage/puts all heaven in a rage' ('Auguries of Innocence'), and asks: 'How do we know but ev'ry Bird, that cuts the airy way, is an immense world of delight, closed by our senses five?' ('The Marriage of Heaven and Hell'). It is not nature – which for Blake as for Swedenborg, is the living face of the divine – but the nature that the new, mechanistic view provided, that troubled him. Indeed, the scientific vision – what Blake called 'single vision and Newton's sleep' – is anti-nature, because it reduces nature to the tedious repetition of eternally whirling wheels, the terrifying 'starry systems' that haunted O.V. de Lubicz Milosz. Blake's phrase 'the dark Satanic mills', which comes from *Milton*, is usually understood to be a reference to the new mills of the Industrial Revolution. But another interpretation sees them as the 'starry wheels', the incessant turning of the Newtonian clock, the great mill of worlds around the pole star that was a sign of cosmic necessity and determinism.[25] These 'dark Satanic mills' grind down human freedom and enslave men's minds with the belief that they are mere cogs in some vast, unstoppable cosmic mechanism. It was with something similar in mind that Alfred North Whitehead once remarked that: 'Life is an offensive, directed against the repetitious mechanism of the universe'. Such a mechanical nature is not alive, but dead.

Blake's reaction to this dead nature, and his belief in the supremacy of the Imagination – again, not simple fancy or novelty, but the creative power behind being itself – could be quite vehement. He took Wordsworth to task for what he saw as the younger poet's idolatry, his veneration of the transient symbol – nature – over its eternal source, the creative mind. Wordsworth and all nature poets were, for Blake, sunk in Maya (illusion, although Blake didn't use this term), entranced by beauty as if by a mirage, and forgetful of the divine energies behind the surface, and within the mind. 'I see in Wordsworth the Natural Man rising up against the Spiritual Man Continually,' Blake wrote, '& then he is no Poet but a Heathen Philosopher at Enmity against all true Poetry or Inspiration'.[26]

Blake's view of nature is similar to Louis Claude de Saint-Martin's. For both, man began as the microcosm, Adam Kadmon, the Great Man, but through some cosmic catastrophe, we fell from the spiritual realms into the world of matter, space, and time and became enslaved by them. Or more precisely, matter, space, time and the Vegetable Glass of Nature are the result of our fall, of our original androgynous spiritual unity fragmenting, shattering, as the vessels shattered after the *tzimtzum*. The 'lifeless universe outside the human Imagination' – Kathleen Raine again – 'is created by the "wrenching apart" of the "eternal mind" resulting, on the one hand, in an externalised "nature" devoid of life, and on the other in a "shrinking" of humanity ('the fallacy of insignificance'?), from the boundless being of Imagination into the "mortal worm" of sixty winters and "seventy inches long", an insignificant part of the externalised nature this wrenching apart has created'.[27] This wrenching apart creates what Blake calls Ulro, the material world, the mundane egg, whose shell obscures the creative energies within it. For Blake, all of creation is human. In a poem to his friend Thomas Butts Blake wrote

> Each grain of Sand
>   Every Stone on the Land
> Each rock & each hill
>   Each fountain & rill
> Each herb & each tree
>   Mountain hill Earth & Sea
> Cloud Meteor & Star
>   Are Men seen from Afar

Blake voiced this theme earlier in the often obscure mythology of *Vala, or the Four Zoas* (1797): 'wherever a grass grows Or a leaf buds The Eternal Man is seen is heard is felt And all his Sorrows till he reassumes his ancient bliss'.[28] As Swedenborg believed, and as the Kabbalists and Hindu mystics believed, the entire cosmos is human, and when man is regenerated, nature will be too.

Waking nature

If he knew of Blake's work, Saint-Martin would have agreed with it.[29] As Saint-Martin's biographer, the occult scholar A.E. Waite, writes, 'The message of Saint-Martin may be fitly termed the

Counsel of the Exile ... It is concerned with man only, the glorious intention of his creation, with his fall, his subsequent bondage, the means of his liberation, and his return to the purpose of his being'.[30] That purpose, we have seen, is to 'repair the universe', which, according to Saint-Martin, is itself, like man, fallen. 'It is in vain that we seek in matter for the real and permanent images of that principle of life from which we are separated unhappily' Saint-Martin tells us.[31] 'The entire universe, notwithstanding all the splendours which it displays before our eyes, can never of itself manifest the truly divine treasures'.[32] It was insights like these that led to Saint-Martin's dictum that we should explain the world by man, not man by the world. 'Man', for Saint-Martin, 'is an exotic plant of the material universe.' 'In his true nature he does not belong to the earth, and the depreciation of his type is the cost of his naturalisation.'[33] Man, for Saint-Martin, is the 'inhabitant of a far country' who has come to earth to fulfil a mission.[34] And in exactly the same way as it is stated in the Hermetic books, Saint-Martin believes that by accepting this mission, man has become a creature of two worlds. He is 'at once great and little, mortal and immortal, ever free in the intellectual, but bound in the physical by laws outside his will; in a word, being a combination of two natures diametrically opposed.'[35] Hence his homesickness, which, Saint-Martin tells us, is the most 'saving sentiment' that we can cherish.

While this places man in an often uncomfortable, even unbearable position, it also makes it possible that by freeing himself, he frees nature. With Blake, Saint-Martin believed that when man regains the status he enjoyed before the fall – when 'he resumes his ancient bliss'; when, that is, his consciousness is once again unified – the fallen world will be redeemed too, will, indeed, be repaired immeasurably. 'The sin of man,' Saint-Martin tells us, 'is the pain of Nature'. 'The universe is on its bed of suffering, and it is for us, O man, to console it.' One way in which man can waken sleeping nature from her trance is through speech. Speech, Saint-Martin tells us, must be restored to the universe. As it is now, the universe is mute, its infinite silence terrifying Pascal and many less articulate souls. But it is only 'he terrestrial man who finds nature silent; for the 'man of desire' – desire, that is, to reunite with his source – 'everything sings in her'. The 'man of desire' must speak truthfully

– say the 'magic word' – in order to liberate sunken nature from its deathlike coma – a reflection that may suggest a different view of the story of Sleeping Beauty. 'We comprehend intuitively that all should speak, and the same intuition instructs us that all should be fluid and diaphanous, that opacity and stagnation are the radical causes of the silence and weariness of nature'.[36] ('Diaphaneity' is a term Gebser uses frequently, indicating a form of consciousness in which the 'ever present origin' is perceived through the various structures in which it is clothed.) Saint-Martin sees that nature is not the reason for man's own sunken condition, but is a victim of it herself, and that it is only man who can draw her up out of the depths. Nature, he tells us, 'is conscious of a secret life circulating through all its veins, and through us as an organ it waits the accents of that speech which sustains it ... It seeks in us the living fire which radiates from that speech ...'[37]

Language and being

Earlier we saw how *kavannot* or 'intentions' were one of the main means of performing *tikkun*, and I gave as one example how speaking thoughtfully and with meaning, rather than our usual empty chatter, is one way of doing this. Saint-Martin agrees. Well in advance of Martin Heidegger, who saw in our incorrigible taste for gossip, for *Gerede* ('talk'), a source of our 'forgetfulness of being', Saint-Martin knew that 'between human conversation and true speech there is an immeasurable distance'.[38] For Heidegger, 'language is the House of Being' (*Letter on Humanism* (1947) and 'Man dwells in this house'. If that is so, then for both Heidegger and Saint-Martin, man has let his house fall into disrepair, in fact, it is even doubtful if it is still inhabitable. Saint-Martin speaks of 'the false and perverted use to which man has put the divine faculty of speech', one of the fundamental abilities that differentiates us from animals, many of whom have systems of communication, but not *language*, with its power to speak of abstract ideas, to use a future tense, to posit possibilities, even to speak falsely. (Can animals lie?[39]) Heidegger believed that our incessant flood of chatter – now facilitated demonically by the miracle of mobile phones – actually alienates us from the 'word', from the true speech that reaches out to authentic being. Being for Heidegger is rooted in speech. in language, which manifests our ability to stand apart from 'the world' and wonder at it and ask questions, which is the essence of thought. (Of all beings we are the one being that asks questions about Being, as Heidegger's mesmerising manner of expression would have it.) Such authentic language Heidegger calls *Rede,* 'speech', as against the empty vocalising of mere 'talk'. Such talk, as George Steiner remarks, 'fosters illusion of understanding without genuine grasp'; as mentioned earlier in a more homely context, saying 'thank you' and meaning it is not as easy as we may think.[40] This talk, Steiner says 'transpires in an echo chamber of incessant, vacant loquacity, of pseudocommunication that knows nothing of its cognates which are, or ought to be, "communion" and "community"'.[41] Saint-Martin agrees. 'So long as we converse,' he tells us, 'either with ourselves or others, only concerning the things of the world [we could say 'life world'], we are acting against the word ... because we are stooping to the world and naturalising ourselves with that which is void of true speech.'[42] Saint-Martin also agrees with Rilke when he says that for 'the man of desire' 'everything sings' in being. '*Gesang ist Dasein*' Rilke wrote in the first of his 'Sonnets to Orpheus' (1922), 'singing is being'. 'We have fallen under the rule of the dead word,' Saint-Martin tells us, and to 'advance toward reality and life we must part somehow with this enormous concourse of rank, empty, earthly and false verbiage corrupting the atmosphere of the mind'.[43] To which one can only say 'Amen!' By doing this we can fulfil the command that 'the transfiguration of man's environment must proceed side by side with the transfiguration of man'.[44] As Saint-Martin said, proof of our regeneration is that we regenerate everything around us.

Freeing nature

I should mention that one of the myths that have gathered around our idea of nature is that it is 'free', and as we've seen, both Blake and Saint-Martin went to great pains to disabuse their readers of this, and to show that, on the contrary, nature is actually enslaved.

'Free nature' is an illusion fostered by the discomfort of self-consciousness, a price we pay for being human. Years ago, Bob Dylan asked the plaintive musical question: 'if dogs run free, then why not we?'[45] The answer to this is that dogs *don't* run free. Dogs, like all animals, are laden with the chains of their biology and are hemmed in by systems of instincts they are unable to throw off. But because dogs aren't expected to conform to the requirements of civilised behaviour, they can engage in activities that to us seem free but really aren't. No dog, or any other animal for that matter, can choose to react to a stimulus or not, in the way that we can. (And whether we do or not is a different question.) In this sense the most harried nine-to-fiver is categorically more free than any dog. As Max Scheler said, animals always say 'Yes' to reality, even when in pain and trying to avoid it. This 'yes' isn't born of a conscious affirmation – they are not 'yea-sayers' as Nietzsche's Zarathustra is – but is rooted in an inability to say 'no', to stand back from life, put it on hold and *decide* whether to accept it or not. If medieval or earlier human beings were, as Owen Barfield believed, 'embedded' in the world, animals are even more so.

What we envy about dogs and other animals is not their freedom, which is severely limited, as is that of every other part of nature, but their lack of self-consciousness, their lack of an ego and its own debilitating concerns. The plunge into 'animal consciousness' sought by some is motivated, I believe, by a desire to escape the human birthright of self-consciousness, the awareness that paralyses Hamlet, but also allows for us to feel a sense of awe and wonder at the world, which, as Heidegger tells us, is the beginning of thought. This predicament was recognised in the nineteenth century by the German playwright Heinrich von Kleist in his remarkable fable 'On the Marionette Theatre' (1810). In it, the narrator expresses his delight at the perfect movements of the dancing puppets he has seen at the theatre, and laments that no human dancer can match such grace. We have it for a time when we are children, but once we become self-conscious – aware of ourselves as independent beings, as the puppets are not – a shadow falls between this early innocence and our actions. Our too acute knowledge of ourselves spoils the unconscious perfection. Our envy of animals, and other illusory manifestations of 'free nature', is rooted in the awkwardness that accompanies awareness of ourselves, and this awkwardness is, as readers may have noticed, another manifestation of our being 'dwellers on two worlds'. Kleist, however, knew that in order to regain this grace, man could only move forward, into a new innocence, a harmony that escapes us now but which he believes it is possible to achieve. 'Grace returns,' Kleist's narrator tells us, 'when knowledge has as it were gone through an infinity. Grace appears most purely in that human form which either has no consciousness or an infinite consciousness. That is, in the puppet or in the god.'[46]

This second grace is available to us, not by sinking back into sleeping nature, which, in any event, we really cannot do, although we make many attempts to do so, but by pushing forward into full awareness of our own creative power. We really have no choice; it is the only direction open to us. Berdyaev agrees with Saint-Martin, Blake and many others when he says that 'creativeness more than anything else is reminiscent of man's vocation before the Fall' and he echoes Kleist when he remarks on 'the union of grace and freedom which we find in creativeness'.[47] (Maslow recorded many incidences when his subjects had peak experiences during some creative act; for example, a student who worked his way through university by performing as a jazz drummer. The student told Maslow of one night when, at two in the morning, he found himself drumming so perfectly that he couldn't make a mistake. He had regained Kleist's 'grace'.) Such creativeness, and a creative attitude to life as a whole, Berdyaev says, 'is not man's right, it is his duty'.[48] 'Man,' Berdyaev writes, 'is a new departure in nature'; he is not a 'fragmentary part of the world, but contains the whole riddle of the universe and the solution to it'.[49] Man is the 'enigmatic being which, though a part of nature, cannot be explained in terms of nature.'[50] He is, Berdyaev says, 'a continual protest against reality'.[51] That protest is made by living creatively.

Living nature

How would a nature liberated from its enslavement, wakened from its trance, appear? Although the idea of a living, conscious nature goes back to the ancients, it enjoyed a popular revival in the 1970s with the publication of *The Secret Life of Plants* (1974) by Peter Tompkins and Christopher Bird. Central to the book was the work of Cleve Backster, an interrogation specialist with years of experience working with lie detectors. One day in 1966 Backster had the idea of attaching his polygraph to one of his rubber plants. Backster wanted to chart the plant's absorption of water by registering its conductivity. As he conducted his experiment and had little success, he wondered at the reading he was getting, and felt it was strangely similar to a reading from a human subject, who was trying to withhold information. Backster wondered what he could do to confirm this, and the idea of posing a threat to the plant came to him, much as an uncomfortable question poses a threat to a human subject, thus triggering a typical reaction. Backster *thought* that he would burn one of the plant's leaves, and at this the polygraph responded wildly. Backster had not actually burnt the leaf, he merely thought about doing it. Yet the plant had reacted. Backster couldn't believe the plant had actually responded, and he repeated the experiment. Sure enough, it did. Backster conducted further experiments, all of which confirmed this initial discovery. Backster had chanced upon the astonishing fact that somehow, plants are able to 'read our minds'. And not only that: they are able to communicate with each other, and are aware of the life around them, animal, vegetable, human, or microscopic. In further experiments with a variety of plants, Backster confirmed Blake's insight that 'a robin redbreast in a cage/puts all heaven in a rage'. At one point the polygraph reacted when Backster poured boiling water down the drain. He couldn't understand why it should, until he realised that the plant was reacting to the death of the microorganisms in the sink. Backster even discovered that plants reacted to human sex.[52]

The details of Backster's experiments and discoveries are fascinating and the interested reader is advised to go to Tompkins and Bird's book, which has acquired the deserved status of a classic. But one of the most important ideas to emerge from Backster's work is his notion of 'primary perception', a kind of universal ur-sense shared by all living things, but which we are oblivious to. When I read this in *The Secret Life of Plants* I immediately thought of Bergson's belief that our brain and nervous system are essentially *eliminative*: that is, they are designed to keep information and stimuli *out* of consciousness. As Tompkins and Bird write: 'the five senses in humans might be limiting factors, overlaying some kind of primary perception, possibly common to all nature'.[53] And as I say in Chapter 2, this makes perfect sense, as in order for us to function in the world, we need to edit out most of the 'interdependent connections' linking everything to everything else. (Again, our 'detachment' from the medieval tapestry with Petrarch's ascent of Mount Ventoux makes the same point: if we were to develop our independent ego-consciousness, we needed to 'lose' our connections with the participatory world.) Yet although we are not immediately aware of this 'primary perception', it is still at work in us, and it may account for some phenomena we call 'paranormal'. In a fascinating paper the consciousness researcher and parapsychologist David Luke suggests that 'our physiology supersedes our cognition in the reception of psychic information ... One interpretation of this is that we may all be continuously psychic, albeit subtly, and yet we remain consciously unaware of the fact, even though our body reacts on our behalf'.[54] It may be the case, then, that our body reacts to what it perceives in 'primary perception', but *we* – our conscious selves – are unaware of this. And again, that we are unaware of most of the 'psychic information' that reaches us makes sense. As Tompkins and Bird write: 'if everyone were to be in everyone else's mind all the time it would be chaos'.[55]

The primal plant

Tompkins and Bird's book draws on many sources for its argument, but two in particular stand out for our purposes here. Earlier I mentioned Goethe as a founder of the Romantic Movement. Johann Wolfgang von Goethe (1749–1832) was one those embarrassingly multi-talented Enlightenment characters that are no longer made today, and along with making breakthroughs in literature, he was also a scientist. In *The Metamorphosis of Plants* (1790) Goethe developed an idea that had been with him for some time, and had come to a kind of climax three years earlier, while on a trip to Palermo, in Sicily. Goethe believed in the existence of what he called the *Urpflanze* or 'primal plant', a living archetype from which all plant forms emerge. 'The Primal Plant,' Goethe wrote, while visiting Palermo's Botanical Garden 'is going to be the strangest creature in the world, for which Nature herself shall envy me. With this model and the key to it, it will be possible to go on for ever inventing plants and know that their existence is logical; that is to say, if they do not actually exist, they could, for they are not the shadowy phantoms of a vain imagination, but posses an inner necessity and truth'.[56] Goethe's 'primal plant', then, is in many ways similar to what Blake above calls a plant's 'Eternal Image and Individuality', its imaginative archetype. Goethe came to his discovery of the primal plant through an intense *participation* in the observation of nature. With Blake Goethe believed that truth is 'a revelation emerging at the point where the inner world of man meets external reality', Blake's 'sun' whose light depends on 'the organ which beholds it'.[57] Goethe showed this to be the case, at least to his own satisfaction, when, after long hours of patiently observing the growth of a plant, he *saw* the primal plant. He did not imagine it – in the negative sense – or merely think it, but actually perceived it. He did not observe his plant though, with the cool detachment of the ordinary scientist, careful to exclude any subjective contribution of his own, but with the imaginative warmth and ardour of an artist – with, in fact, the kind of phenomenological 'love' that Max Scheler believed was the key to discovering any truth worth knowing. Goethe's 'imaginative observation' revealed the plant to be a living *idea*, a 'soul' that revealed itself to the lovingly attentive eye in the process of its development.[58] Goethe established that nature was alive, and that if we are unaware of this, it is through the passivity of our own perceptions. If our consciousness was more active – participatory, regenerated – then nature would appear to be more alive too. It was on Goethe's notion of a kind of 'active perceiving' that Rudolf Steiner would base his ideas about what he called 'spiritual science'. Through Goethe's revelation of the *Urpflanze*, Steiner was led to the insight that 'Man is not there in order to form for himself a picture of the finished world; nay, he himself cooperates in bringing the world into existence', an insight we have already come across in several other thinkers.[59]

The angel of the Earth

Most of us are aware of the notion of the earth as a living being, as 'Gaia', through the work of James Lovelock. Yet Lovelock himself is at pains to point out that the idea of a 'living earth' is for him really only a metaphor, an *'aide pensée'*, an aid to thought, and not to be taken literally.[60] Lovelock's concern for the health of this *aide pensée*, however, has led him to make some radical statements. He speaks of human beings – for him, the clearest threat to 'Gaia' – as 'a plague of people', an animal who had 'begun the demolition of the Earth', the species equivalent of ... Jekyll and Hyde', and a 'planetary disease', and advises that our work to help Gaia should be carried out 'as if we were about to be attacked by a powerful enemy', a troubling phrase that suggests the possibility of some kind of 'pre-emptive strike' against the source of the problem – ourselves.[61] Yet an earlier nature-friendly thinker who took the idea of the earth as a living being seriously, not simply as a metaphor, and who first proposed that plants had an inner life, harboured no such negative thoughts about people. For him they are the earth's sense organs, its means of becoming conscious of itself, an idea voiced more than once earlier in this book. The earth itself, according to him, was, in the words of his great admirer William James, our 'great common guardian angel, who watched over all our interests combined', a remark that echoes Vladimir Solovyov's idea, mentioned in Chapter 3, that the universe 'from time immemorial has been interested in the preservation, development, and perpetuation of all that is really necessary and desirable for us'.[62] Not an idea, I think, that would go over well with James Lovelock.

Gustav Fechner (1801–1887), who I am speaking of above, was an amazingly prolific and dedicated German scholar and scientist, who spent practically all his life in the University of Leipzig. He is known as the founder of 'psychophysics', the discipline that formed the basis for scientific psychology, and Fechner's research into the relationship between psychological sensation and the intensity of physical stimuli is still considered important. Freud and Einstein are among the many people influenced by his work.[63] But Fechner was a remarkably versatile individual, and along with tutoring and translating innumerable scientific works in order to support himself, he also carried on a separate career as 'Dr Mises', a pseudonym he used for a series of humorous, satiric, and philosophical essays and wise sayings, that are often as insightful and elegant as those of his countryman Georg Christoph Lichtenberg. But what is important to us about Fechner's work – and the reason he is included in Tompkins and Bird's book – is the insight – literally – that came to him after he went through what the historian of psychology Henri Ellenberger calls a 'creative illness'.

In 1840, at the age of thirty-nine, overwork and eyestrain sent Fechner into a kind of nervous collapse. From looking at the sun in order to observe the phenomenon of after-images, he became almost blind, and had to stay in darkened rooms, his eyes bandaged because any exposure to light caused him insufferable pain. He couldn't read and this led to depression and an inner disorientation, a perpetual sense of confusion, with his thoughts running out of control, something Swedenborg, who also endured a 'creative illness', experienced.[64] Fechner mistrusted doctors but tried various 'alternative' remedies, which only made things worse. His body could absorb no food and he shrank to skin and bones. He also found it almost impossible to speak, every attempt throwing him into fits of distress.

An acquaintance heard of his plight, and dreamt that she had prepared some food that he could eat. She visited him and offered her cure. It worked for a time but did not prove decisive. After three years of this torture, Fechner, in sheer anger, began to speak loudly, ignoring the pain. This seemed to work and his voice improved. He then decided to expose himself to the light suddenly, and not gradually, as he had tried before. The effect was astonishing. Visiting his garden in full daylight, Fechner pulled off his bandages and felt that he had been given 'a glimpse beyond the boundary of human experience'. 'Every flower', he wrote, 'beamed upon me with a peculiar clarity, as though into the outer light it was casting its own. To me the whole garden seemed transfigured, as though it was *not I but nature that had just risen up again*' [my italics.] 'And I thought: so nothing is needed but to open the eyes afresh, and with that old nature is made young again. Indeed, one will hardly believe how new and vivid is the nature that meets the man who comes to meet it with new eyes.'[65]

Along with this vision of the flowers as living, conscious beings, and not mere automata, a new optimism and confidence filled Fechner, a sense of destiny, of having an important message for mankind, something common to individuals who have gone through a 'creative illness'. Fechner had begun as an atheist, but a reading of the *Naturphilosophie* of Friedrich von Schelling had cured him of that, and before his illness he had written a short devotional text, *A Little Book of Life After Death* (1836) in which he argued that what we take as waking life is really only a half-sleep – something Gurdjieff would have agreed with – and that in death we enter full consciousness. What emerged from Fechner's experience was the realisation that nature, the whole universe in fact, is alive and conscious. Fechner expressed his ideas about the inner life of plants in his book *Nanna: The Soul Life of Plants* (1848) and in *Concerning Souls* (1861) he broadens this insight to cover the entire cosmos. In a sense, in order to see, Fechner had to become blind, and he would say that we, who still regard the universe as a lifeless machine, are the truly blind ones. Fechner thought in terms of a 'continuous, perpetual creation', and his experience allowed him to share in the 'primary perception' that Cleve Backster believed was enjoyed by all nature. Fechner's basic theme is that, just as we have an 'inside', an inner, 'soul world', so too does the universe, and it is only our blindness that prevents us from seeing this. The earth, stars, and galaxies – which Fechner considered angels – as well as a pebble on the road, is shot through with life and consciousness, not the kind of acute self-consciousness we experience, but the kind of diffuse consciousness Backster's plants seemed to have exhibited. (In his 1961 novel *Solaris*, about the discovery of a living planet, the Polish science-fiction writer Stanislaw Lem (1921–2006) paid homage to Fechner by naming one of the characters after him. In 1972 the novel was made into a disturbingly beautiful film by the Russian filmmaker Andrei Tarkovsky.)

Fechner believed that 'life was primordial and organic matter prior to the unorganic', the complete opposite of the standard scientific view, which sees life as an accidental product of matter.[66] This was a view Fechner shared with later 'vital' philosophers like Bergson and Whitehead, and the contemporary philosopher of mind David Chalmers seems to agree with Fechner that consciousness pervades the universe.[67] As William James said, for Fechner 'original sin' is 'our inveterate habit of regarding the spiritual not as the rule but as an exception'.[68] He contrasted what he called the 'day view' that regards the 'whole universe' as 'everywhere alive and conscious', with the 'night view', that sees life as an anomaly in a dead, mechanical cosmos. For his contemporaries, 'in the midst of nature as a whole ... animated creatures appear like islands in the universal ocean of dead and soulless bodies.' This was false, but understandable, since 'it is in a dark and cold world we sit, if we will not open the inward eyes of the spirit to the inward flame of nature'.[69]

Fechner had opened his inward eyes and he saw. 'The real and vital question,' he wrote, ' is, not whether the world has a spirit about which we alone know, but whether it is Spirit which knows about us'.[70] Fechner, however, did not believe that matter was unreal, as idealist philosophers did; and as we've seen, he also didn't believe that matter was the only real thing. What was real was the union of matter and spirit. In much the same way that for an idea or value to be actualised it must have form and be rooted in reality, Fechner believed that for spirit to be perceived it must be embodied. Fechner argued by analogy. Just as to the totality of all we see or perceive in man there corresponds a centre, a soul, that we do not see – with the physical eye – but perceive, as Scheler believed, intuitively, so too to the totality of all we perceive in nature there corresponds a unified Being which we may assume is only immediately apparent to itself. And just as we cannot find the human soul with a microscope – something Swedenborg, a fellow scientist, discovered – we cannot find the soul of the earth or the cosmos with 'telescopes, deep drilling, or measurements ...'[71] We can find it, however by opening the inward eye which, as Fechner discovered, makes 'old nature young again', and all creation 'new' and 'vivid'. As Goethe did when he discovered the primal plant, Fechner saw that for nature to be alive, we must be too.

Let us take a look then at what that might mean.

6. The Participatory Universe

As readers of this book may have concluded by now, my ideas about consciousness and its relation to the world, as well as those of most of the thinkers I've discussed, are not those most commonly held by modern science. The basic idea about consciousness from the scientific point of view – acknowledging of course that this is a generalisation and that scientists quibble among themselves about the details – is that it somehow emerged from matter, and that, as mentioned earlier, quoting Francis Crick, our joys, sorrows, memories, ambitions, sense of personal identity and free will – not to mention everything else in our heads – are in fact 'no more than the behaviour of a vast assembly of nerve cells and their associated molecules'. According to this account, at some time in the history of life on earth, what we call consciousness simply popped up, squeezed out somehow by organic matter, and through the application of purely accidental forces, developed from some minimal state of proto-awareness into my own consciousness writing this book and yours reading it. Exactly when and where that initial blip of consciousness emerged is unknown, but we must assume it started somewhere. The organic matter out of which consciousness is thought to have emerged was, according to modern scientific accounts, itself a product of chance forces, and so for consciousness to have done the same is in keeping with the whole temper of modern science's explanations for things. In fact, as I remark in Chapter 1, the basic scientific explanation for the universe, life, and consciousness is that 'it just happened'. This conclusion is supported by years of research, by hugely expensive experimental equipment, by stacks of papers and by well-honed arguments. But in the end I think my précis of it is unassailable. Give us the fact of the Big Bang scientists say, allow us this, and the rest follows. But if you say 'All right, but please tell me why the Big Bang happened?', then the scientists will either say that you cannot ask that question – it is not 'scientific' – or, as Stephen Hawking does, that it emerged from a 'quantum fluctuation in a pre-existing vacuum', which, as I remark in Chapter 1, is a very abstract way of saying that it came out of a ripple in nothing. Or, more briefly, that it just happened.

The point here is that no one is in doubt that 'life, the universe, and everything' – as Douglas Adams phrased it – happened. We want to know why it happened. And by 'why' we do not mean 'through what agency' but 'for what purpose'. And this is something science cannot tell us. 'Because' may be an answer for children, but not for men and women who want to know why they exist.

For all the rhetoric about closing in on the secrets of the universe, about capturing the 'God particle' and other dramatic announcements, no one, not Stephen Hawking, not Steven Weinberg, nor any other scientist, knows why the Big Bang happened. They don't and neither do I. It would be refreshing if, instead of obfuscating this fact with rhetoric aimed at blinding us with science, someone would simply admit this. Some scientists do, but they are generally presented as 'mavericks' or 'outsiders' or 'pessimists'. This last tag is important, because, for all its fear of science – witness our concerns over GM crops, nuclear energy, or the MMR vaccine – the public believes in science – Big Science – and wants it to give it directions, wants to know that it *has* answered the Big Questions – or that it soon will – so we can be done with them and just get on with life. But very rarely do scientists themselves explain how tentative their assessments of things are, or admit the uncertainty at the root of the ideas on which they hope to base a 'theory of everything'.

The reasons for this are many and complex. Some have to do with keeping things upbeat, in order to secure funding and have public support. Some have to do with avoiding the opprobrium of their colleagues, with not losing their scientific 'street cred', and appearing 'soft'. And some have to do with the fear that, if science was shown to be less positive, less secure, then the bogey of God, religion, and the 'bad old days' would return. Hence the attention some atheists have generated in their efforts to broadcast their views to the public. Scientists who have braved these concerns and 'blown the whistle' generally meet with derision from their fellows and are relegated to the lunatic fringe. Witness the biologist Rupert

Sheldrake and his important book *The Science Delusion* (2012), in which he argues – convincingly, to my mind – that much modern science lives on 'promissory notes'.

The phrase 'promissory materialism' comes from the philosopher of science Karl Popper, and by using it Sheldrake questions science's tactic of assuring us that, although its strict reductive materialist approach may not have all the answers now, it will eventually. It will, science tells us, because it is the *only* approach available to us, the only approach to knowledge and the only approach to truth. Everything else is pseudoscience or imagination or mysticism; at any rate, something less 'real' than what they consider science. Sheldrake's argument is that this kind of attitude is not true science, which is based on scepticism – that is, on questioning premises and keeping an open mind – but dogma, in the same way that the religious fundamentalism that science wants to save us from is based on dogma. This is not science but scientism. But the followers of scientism still hold most of the cards, and so they can say that what Sheldrake is doing isn't science, but cranky pseudoscience. And although the public is interested in this debate, any support for Sheldrake's or any similar arguments is easily slotted into the 'woolly new age' basket and left there. We may be getting closer to the 'paradigm shift' more open-minded thinkers have been talking about for decades, and Sheldrake's work has certainly helped us reach this point. But we're not there yet.

Some loose strings

Yet, if you do a bit of research, you discover that many of the certainties on which scientism is based are shaky indeed. For example, one of the central ideas in modern cosmology is 'string theory', a theoretical model that aims to unify quantum mechanics and general relativity, two giants of modern physics that have hitherto refused to be wed. Relativity and quantum mechanics are both 'right' but mutually exclusive. On the macro level of planets, stars, and galaxies, relativity reigns but down with the particles quantum is king, and their antipathy to each other creates a kind of cosmic crisis, at least among physicists. Tying the knot between them with string theory will, many scientists believe, finally wrap the universe up into a neat bundle, with no loose ends. String theory argues that, rather than elementary particles being dimensionless, they are actually one-dimensional oscillating lines or 'strings' – an image that, for what it's worth, suggests to my mind both Pythagoras' octaves and the old occult notion that reality is made up of different levels of vibrations. (Remembering my earlier remarks about dark matter, it does seem that in some ways, contemporary science is finding its way to some old occult truths.)

String theory and its cousin 'superstring theory' are both good contenders for the still elusive 'theory of everything', science's big push to finally explain it all. Yet according to some scientists, string theory is 'not even wrong'. This is the title of a book by Peter Woit – *Not Even Wrong: The Failure of String Theory and the Search for Unity in Physical Law* (2007) – showing how string theory fails to meet Karl Popper's 'falsifiabilty test' for something to even qualify as a scientific theory. This is a challenge the strings have also had to face in Lee Smolin's *The Trouble With Physics: The Rise of String Theory, the Fall of a Science, and What Comes Next* (2007). Both books argue that string theory makes no predictions and cannot be tested, begging the question whether it is a theory at all, and not a form of scientific theology. Nevertheless, it is a fundamental tenet of the 'standard model' and commands near absolute obeisance in physics departments. Commenting on this situation, George Steiner remarks that regarding string theory, 'So far, not a single new testable prediction has been made; not a single genuine theoretical puzzle has been solved', and asks 'If string theory is immune to either proof or cancellation, if it is nothing but a mathematical game of great beauty and arbitrary license ("it can be made to mean anything" comments one ranking cosmologist) then is it a science?'[1]

Another glitch in the 'theory of everything' game is a problem known as Gödel's Theorem. Kurt Gödel (1906–1978) was an Austrian American philosopher and mathematician, and he is famous for his two 'incompleteness theorems'. The essence of these is that in any closed system, there are propositions that cannot be proved or disproved using axioms within that system. To be proved or disproved, they must refer to axioms *outside* the system. With Gödel's theorems there are no absolutely closed systems, which, it seems, would make a 'theory of everything' impossible, or at least turn it into something like those strange, recursive, self-referential works of M.C. Escher, with hands drawing each other or stairs returning on themselves. A theory of everything would have to include the theory itself as well as the theoriser and so on ad infinitum and would seem to enter the oddly unreal territories of reiteration, such as we find in fractals. I am no physicist but my own belief is that after superstrings, we will no doubt find something else, as with each of its secrets we uncover, the universe will add another one, perpetually pulling rabbits out of its hat, or, like a novel we are reading while it is still a work in progress, adding chapters as we go along.² A 'theory of everything' would also seem to meet Wittgenstein's definition of the 'mystical', and I somehow doubt that Hawking and the others in pursuit of this particular holy grail have this in mind . In the *Tractatus Logico-Philosophicus* (1921) Wittgenstein writes: 'Feeling the world as a limited whole – it is this that is mystical'.³ How else, in a 'theory of everything', would the world be perceived except as a 'limited whole', although, as mentioned earlier, this leads to the logical problem of *who* would perceive it as such, as in order to perceive it as a whole, a perceiver would have to be *outside* it, and if he were, then it could not be a 'whole' or 'everything'? Wittgenstein also wrote that 'even when *all possible* scientific questions have been answered' – which, one imagines, is what a 'theory of everything' would want to do – 'the problems of life remain completely untouched.'⁴ So perhaps we should leave it at that.

Consciousness explained?

There are quite a few other chinks in scientism's armour, but the main ones that concern us are those regarding consciousness.⁵ I've already mentioned Francis Crick's belief that everything passing through my mind as I write these words and everything in yours as you read them is ultimately nothing but 'the behaviour of nerve cells'. But Crick at least assumes there is something called consciousness, something others in the 'explaining consciousness' game are less generous about. The philosopher Daniel Dennett, for example, in his highly influential *Consciousness Explained* (1992) is much less charitable. Dennett's five hundred plus densely written pages are dedicated to the proposition that consciousness doesn't really exist, that it is an illusion produced by the brain, and that we are all really robot zombies, who only *think* we are conscious. But as I've asked in a previous book, if that is the case, then *who* does Dennett want to explain consciousness to? Perhaps Nicholas Humphrey, who, like the rest of us, only thinks he is conscious, may know, as he too is eager to nail consciousness down. In *Soul Searching* (1995) and other works, Humphrey is at pains to lay the ghost of consciousness to rest, an aim driven perhaps by something more than scientific zeal. As he admitted at one point, 'unexplained subjective experience arouses an irritation in me', a sentiment that ranks with James Lovelock's call for anti-human 'pre-emptive strikes' (Chapter 4) as high on my worry list.[6] As a scientist, Humphrey is determined to reject anything that smacks of the supernatural, and as consciousness as we usually understand it is immaterial, that is, not perceivable by the senses, it is fair game for this ghost buster. Another thinker with consciousness in his sights is the philosopher John Searle, author of *The Mystery of Consciousness* (1990) and other works. While critical of Dennett and other 'explainers' who want to eliminate mental phenomena, Searle has no hesitation in reducing consciousness to an 'ordinary biological phenomenon comparable to growth, digestion, or the secretion of bile', a remark that puts him the same league as the eighteenth century French physiologist Pierre Cabanis, who first declared, as Searle does, that the brain secretes thoughts as the liver secretes bile.[7] Not only would this idea of equating thinking with other bodily functions not be amenable to Nikolai Berdyaev, but it would also mean that my grappling with Searle's thoughts is on a par with digesting my lunch. Like Nicholas Humphrey, Searle is eager to 'demystify' consciousness, to take it down a peg, and to turn the 'mystery of consciousness' into the 'problem of consciousness', a problem that science, and its ally reductive philosophy, will no doubt soon solve – another example of Sheldrake's 'promissory notes'. To this end Searle is almost religious in his repeated assertion that the 'brain causes consciousness'. We need, Searle tells us repeatedly, to explain 'exactly how neurobiological processes in the brain *cause* our subjective states and awareness'.

A hard problem

But that, as they used to say, is the $64,000 question. And no one has the answer. Just as no one has the slightest idea why the Big Bang happened – if, indeed, it did – no one knows how or even *if* 'neurobiological processes *cause* our subjective states and awareness.' No one knows this, even though huge, difficult, and highly regarded books by eminent scientists and philosophers argue that pretty soon they will. No one does. It is, as is the case with other pursuits of 'promissory science', an assumption. Yet, one question that immediately comes to mind is whether we need a brain at all to be conscious. As I write in *A Secret History of Consciousness*, there are cases on record in which individuals who led highly active, fruitful, and intelligent lives, did so apparently with no, or extremely little, brains.[8] Yet even leaving aside these difficult and frankly bizarre cases, the mysterious problem of consciousness *itself* raises sufficient obstacles to the kind of explanation people like Dennett, Humphrey, Searle, Crick and others are looking for. There is, for example, the 'hard problem' which wedges open what is known as the 'explanatory gap'.

The 'hard problem' is easy to state but difficult to answer. It is this. No one knows how firing neurons become subjective experience, the kind that irritates Nicholas Humphrey. No one knows how Crick's molecules become curiously intangible but highly meaningful things such as our joys, sorrows, memories, as well as Crick's thoughts about these. No one knows. No one has ever seen a neuron turn into a thought or a molecule become an idea, or even anything remotely like that. I haven't and I don't believe anyone will. And this is because neurons, brain cells, and molecules are one kind of thing and ideas, thoughts, feelings, beliefs are another. I don't believe that my brain 'causes' my consciousness just as I don't believe that my television set or radio 'cause' the programs I watch or listen to with them. If I took apart my television set and examined every molecule of it, I would never find the 'cause' of the program I had just been watching. If I smashed my skull in I would certainly upset my consciousness, just as if I smashed my television set I would be unable to watch what was on – something, I must admit, I am often tempted to do. But while my TV would be broken, the program I was watching would be unaffected, and would still be running on millions of other TVs. I believe that my brain acts in some way as a radio or television receiver. It doesn't 'cause' the consciousness I have any more than my TV or radio 'causes' what is broadcast on them. My TV or radio *tunes in* to one program by filtering out the others. They are eliminative in the same way that Bergson believed our brains and nervous systems are. Rather than *cause* my consciousness, my brain *limits* the amount of consciousness I experience. In this sense, consciousness is more like a radio wave than bile. Years ago the novelist Upton Sinclair wrote a book about telepathy called *Mental Radio* (1930). One wonders if Searle is working on *Mental Bile*.

Explanatory gaps

So much for the hard problem which is a problem only if you are obsessed with reducing immaterial things like thoughts, ideas, feelings, and values, to material things like neurons, nerve cells, and molecules. The 'explanatory gap' is the yawning, impassable chasm between these two kinds of things.[9] It is the unbridgeable gulf between apples and oranges or chalk and cheese, or, more precisely, between the refraction of electromagnetic radiation that constitutes a sunset and the mystical awe I feel looking at it. But it is only a gap for those who want to explain oranges in terms of apples or cheese in terms of chalk or awe in terms of refraction. And this is not a question only woolly metaphysicians are interested in.

According to Stuart Kauffman, a renowned theoretical biologist and complex-systems researcher who has done pioneering work in self-organisation – the process by which order arises out of disorder, or a cosmos out of chaos – 'nobody has the faintest idea what consciousness is ... I don't have any idea. Nor does anybody else, including the philosophers of mind.'[10] The German neuroscientist Wolf Singer is equally doubtful about the kind of equation between consciousness and brains that excites John Searle. 'We encounter extreme difficulties', Singer writes, 'when we attempt to explain how exactly the *qualia* of our subjective experience emerge from neuronal interactions'.[11] As mentioned, *qualia* is a Latin term used to refer to inner experiences such as pain, taste, or the delight we feel in perceiving colour; it refers to the 'value content' of mental experience, the difference between electromagnetic radiation with a wavelength of 650 nm and the colour red. Particular neurons may be involved in, say, my appreciation of a glass of wine. But the neurons and the appreciation are not the same thing, just as the words on this page and your understanding of them are linked, but different.

Nonlocal neurons

Singer's work has focused on another central question of consciousness, what is known as the 'binding problem'. How, out of the mass of sensory and other data impinging on it – what William James called 'the blooming, buzzing confusion' – does the brain form a whole, a comprehensible unity, that we experience as 'the world'? Singer discovered that the answer to the 'binding problem' may be found in the synchronised firing of neurons in different parts of the brain. The neurons in question are not in *physical* contact with each other, but temporal. They are separated spatially but are linked by time. As the pioneering neurophysiologist Charles Sherrington said, 'Pure conjunction in time, without necessarily cerebral conjunction [meaning not next to each other in the brain] lies at the root of the problem of the unity of mind'.[12]

Yet how do the different neurons, located in different areas of the brain, 'know' when to fire so that they are all in synch? How do they know when to act in unison? Physical contiguity would explain an electro-chemical 'message' passing from one neuron to another, but the neurons in question are not contiguous. The effect is rather as if a hundred or so people, scattered across the globe and without contact with each other, all knew when to perform some action simultaneously, which would produce a co-ordinated meaningful result. In some ways Singer's idea is reminiscent of what is known as 'quantum non-locality', the fact that under some conditions elementary particles not in contact with each other and not subject to a connecting force can be correlated without any transfer of energy or passage of time – that is, simultaneously and at a distance. (Einstein didn't like this idea of some 'spooky action at a distance', but failed to disprove it. [13]) The idea also seems to have some resonance with 'synchronicity', the 'acausal connecting principle' that C.G. Jung posited as responsible for the phenomena of 'meaningful coincidence', those strange episodes when there is a direct but unexpected correlation between our inner and outer worlds, a connection not of cause and effect but of meaning.[14]

Such ideas usually elicit an impatient 'harumph' from most scientists, but Singer is open to the idea of what we usually call the paranormal. Speaking of telepathy – another form of 'spooky action at a distance' – and other such oddities, Singer says that: 'If only one of these parapsychological phenomena finds a convincing scientific validation, then we have to admit that our brain theories miss a really essential ingredient ... It is at least theoretically conceivable that there is a class of phenomena which are characterised by the fact that they are not repeatable ... In this case our scientific approach is bound to fail'.[15] The philosopher David Chalmers, who has in many ways widened the 'explanatory gap' – and who John Searle subjects to a particularly robust drubbing – believes that no matter how much we learn about the brain, purely physical or materialist theories will never reveal the secret of consciousness. For this belief, Chalmers and his like minded colleagues have been dubbed 'mysterians' by their more optimistic opponents, as they want to hold on the 'mystery' of consciousness and not, like Searle, to turn it into a 'problem'. Chalmers' own approach is much more in line with the ideas of Bergson and Whitehead, who both posit a kind of 'pan-psychism', in which *everything* in some way participates in consciousness, an idea we met with in the last chapter in Cleve Backster's 'primary perception'.[16] It may be that Iain McGilchrist, an expert in neuroimaging, sums up the situation best. 'The fundamental problem in explaining the experience of consciousness,' he writes' 'is that there is nothing else remotely like it to compare it with.' This would, I think, exclude bile.

Consciousness evolves

Given that the kind of definite explanation of consciousness sought for by many scientists and philosophers does not seem to be forthcoming – or at least is as forthcoming as a 'theory of everything' – and that many eminent scientists and philosophers are actually doubtful that it will ever be found, and given that scientists such as Wolf Singer, a director at the Max Planck Institute for Brain Research in Frankfurt, is open-minded about parapsychology, we can, I think, entertain some ideas about consciousness that are off the beaten scientific track. One of these is the evolution of consciousness, something we touched on in the last chapter.

What I mean by the evolution of consciousness has nothing to do with Darwin's 'dangerous idea'. I do not mean how consciousness evolved by chance out of 'un' or 'non-consciousness', or how our chance consciousness evolved, under a variety of environmental pressures, from some dim, vague awareness to our own acute sense of self and the world. As we've seen, in these and other mainstream ideas about consciousness, it is ultimately a chance outcome, an epiphenomenon, of some physical or material reality. The evolution of consciousness I am thinking of takes consciousness as fundamental and irreducible and not solely localised inside our heads. Consciousness from this perspective did not emerge out of matter at some time in the past. Consciousness was there to begin with, and it would be more correct from this perspective to say that matter emerged out of it, a point we will return to.

Consciousness is the bottom line reality. It is, I accept, a mystery, but it is one I have an immediate, direct awareness of – I am, in fact, immersed in it – and it is one that is inseparable from my having any experience at all. Everything I know comes to me through consciousness, from the most intimate personal detail of my life, to my knowledge of what is happening at the furthest fringe of the observable universe, to my questions about the infinite complexities of the human brain. Without consciousness there would be none of this.

But consciousness is not static. It changes over time, both in the individual – me – and in the wider culture. And as consciousness changes, so does the world. As we saw in the last chapter, changes in human consciousness beginning in the Renaissance and carrying through to the eighteenth and nineteenth centuries radically altered our ideas about and relationship to nature. It is only from Husserl's 'natural standpoint' that the world and our consciousness of it seem fixed. The kind of consciousness we find ourselves with, and which we believe is just as 'given' as the world we find ourselves in, has developed out of earlier forms. From this perspective, the people of an earlier time did not live in the same world as we live in, nor did they have the same kind of consciousness as we do, something we looked at in our discussion of medieval tapestry and Renaissance perspective.

Thinkers I have discussed earlier in this book – Bergson, Schwaller de Lubicz, Gebser, and others – have, in different ways, looked at the evolution of consciousness. In this chapter I am going to look at some ideas about that evolution that, I believe, can help us find a way of understanding how we can 'repair' the universe at a fundamental level. Indeed, these ideas can help us do more than repair it. They can help us take up our responsibility of being co-creators of reality. As different thinkers have expressed it in this book, the cosmos is incomplete without our contribution to it. By the end of this chapter I hope I will have shown at least some ways in which we can make that contribution and what a world in which we are aware of making it would be like.

Barfield's participation

Owen Barfield, who we met in the last chapter, first came to the idea of an evolution of consciousness through a study of language. In his first book, *History in English Words* (1926), Barfield wrote that: 'In our own language ... the past history of humanity is spread out in an imperishable map, just as the history of the mineral earth lies embedded in the layers of its outer crust. But there is this difference ... whereas the former can only give us a knowledge of outward dead things, language has preserved for us the inner living history of man's soul. It reveals the evolution of consciousness.'[17] What Barfield meant by an 'evolution of consciousness' is 'the concept of man's self-consciousness as a process in time', an idea, he believes, that sums up the essence of Rudolf Steiner's philosophy.[18] That is to say, it is the process through which we have *arrived* at our present self-conscious. What this means is that, not only have our *ideas* about the world changed over time – the study of this change is known as 'the history of ideas' – but the *consciousness* entertaining those ideas has too. An ancient Greek and a twenty-first century Londoner certainly have very different ideas about the world, but – as we saw in our discussion of the medieval and Renaissance world views – how they would actually see and experience the world would also be different. Barfield came to this insight through a study of metaphor, how 'figurative speech' seemed to make the world come to life, to make it more meaningful. (As we saw in Chapter 2, metaphor is a right brain way of understanding the world, which the left brain simply doesn't 'get', and the right brain is geared toward participating in a living world.)

Barfield recognised that this figurative, 'living' way of speaking, seemed to be something more pervasive in the earlier stages of language, and that, by contrast, contemporary language was much more prosaic; we live, as mentioned earlier, in what the literary philosopher Erich Heller calls 'an age of prose'. Metaphors, Barfield realised, have an initial power to 'shock', to awaken consciousness to new possibilities, to its 'relationality', how something is 'like' something else. This is why we find great literature and poetry so moving. The first person to recognise that a woman's face bloomed or that anger burned must have felt something like a mini-mystical experience. Yet over time, metaphors lose this power. They become, as Emerson remarked, 'fossils', or, as Nietzsche said, like coins whose face has been worn down by use. Our everyday language is full of metaphors we no longer notice, because their use is so widespread and habitual that we take them for granted. So now, as I am trying to capture these ideas so that you can grasp them, I no longer feel any 'shock' at the notion that I can 'capture' an idea or that you can 'grasp' it – two metaphors we use unconsciously.

(Ideas are not animals nor are they a hammer, or anything else I can pick up, except metaphorically.)

In looking at language from an earlier time, Barfield recognised that it could produce this 'shock', just as poetry could, and the reason for this is that it was much more figurative. Just as poetry can induce a sense of the world's hidden possibilities, language from an earlier time can as well, and this was true, Barfield concluded, because the consciousness of an earlier time was different from our own, and this difference is reflected in language. For this earlier consciousness, he saw, the world itself was more living, more 'figurative', and less 'matter of fact' than our own world is. But Barfield also recognised that the history of language seemed to move from the concrete to the abstract. 'Whatever word we hit on,' he wrote, 'if we trace its meaning far enough back, we find it apparently expressive of some tangible ... object or some activity'. 'The movement of meaning,' he saw, 'has been from the concrete to the abstract.'[19] So our abstract concept of 'spirit' as an immaterial 'something' has its roots in the Greek *pneuma* and the Hebrew *ruach*, both of which designate an 'indwelling soul' but also the movement of air.

To language theorists like the Oriental scholar and philologist Max Müller (1823–1900), who was extremely influential in the late nineteenth century, this suggested that early language was made up of very simple literal words and phrases for tangible, sensible objects, and then gradually we began to use these simple words mistakenly, to refer to 'spiritual' realities . According to Müller, at some unspecified time, our ancestors began to use words 'metaphorically', that is, to more or less consciously write 'poetry' while describing their experiences, and through this 'poetry' confused metaphor or 'turns of phrase' with actual but non-material realities. (This was thought of as a kind of 'disease' of language, a belief that carried through into modern times, in a variety of ways: logical positivism, linguistics analysis, deconstructionism.) Barfield even jokes that, in order to account for the sudden rise of metaphor out of its literal roots, Müller has to posit a whole generation of poets, who, out of nowhere, suddenly began to spout wonderful verse, and that they passed this habit on to the next generation and the next. Yet, Barfield thought, this would suggest that we, who have inherited this knack several millennia on, should be speaking poetry with every word, and that the poetry from an earlier time should strike us as less poetic. But Barfield saw that the exact opposite was true: our language is much more prosaic, and, as he found in his studies of earlier language, poetry from an earlier time strikes us as much more poetic.[20] Modern poetry itself is in many ways really little more than truncated prose, a tendency that began in Barfield's time with T.S. Eliot's *The Waste Land* (1922). What was wrong?

The answer, Barfield saw, was that Müller and the language theorists who followed him looked at language from an unquestioned Darwinian perspective. They assumed that, just as simple organisms became more complex over time language too must follow a similar path. It too must 'progress'. So simple 'root' words denoting sensible, tangible things gradually grew into our abstract, highly metaphorical language. Yet, as Barfield saw, the evidence from language simply didn't agree with Müller. It was not the case that early man accidentally hit on the clever knack of speaking figuratively, but that the world he perceived was, as mentioned, actually more figurative, more alive. Early man's consciousness was different from our own, and hence the world he lived in was different too. Müller and those like him made the mistake of projecting how they saw the world back into the consciousness of early man. Because of this, they then had to explain how language, which they assumed started out as clichéd primitive grunts, designating a simple *this* and *that*, became figurative. Hence the prehistoric poets.

Barfield's suggestion was much more interesting and much closer to what language itself tells us. Early language is more figurative, metaphorical, and alive, because the world it was spoken in was too. It was a world much more inclined to the kind of 'picture consciousness' Rudolf Steiner spoke of and which he believed was the kind of consciousness we experienced before the rise of our modern rational consciousness. It was a world, Barfield tells us, in which consciousness participated, that is, it 'lived' with it. It was not like the world that Müller knew and which he projected back into the mind of early man, a world of discreet, well-defined, stable, 'objective', 'external' things, but a much more flowing, fluid, and connected world. (Readers will recall that this is precisely the difference between how the left brain and the right relate to the world.)

The origin of origin

Language, Barfield tells us, did not arise out of an attempt to understand such a world as Müller projected back into history, because such a world did not exist prior to language. It was not as if a consciousness like ours, but without language, suddenly awoke in a world like ours – but without, of course, all our technology – and hit on making sounds as a way of describing it. The polarity of 'nature/language', or 'world/mind' did not arise one after the other but simultaneously. It was not world first, then mind, or nature first, then language. There was no world for language to try to describe because there was no world before language. This is why Barfield says asking about the origin of language is like asking about the origin of origin.[21] One doesn't follow the other. Both are two sides of a whole. It is only we, who have become stranded on one side – the Other Side – who think that the kind of world we perceive must have preceded any attempts to create a language to describe it. We perceive the kind of world we do because of language.

Just as child does not have a world until its consciousness begins to separate itself from everything around it, so too our ancestors did not have a world, did not have nature, in the way that we do, because their consciousness was not as separated from everything around them as ours is. As Max Scheler believed, we do not have to guess about someone else's 'inside'. We know it exists intuitively because at one point our inside and theirs were the same. Barfield extends this to nature and the world. Earlier in our evolution, we were as aware of nature's 'inside' as we were of our own; with it we shared in Cleve Backster's 'primary perception'. Or, more precisely, our inside and nature's was the same; we made little distinction between the two; certainly we did not recognise the kind of rigid difference we do now, with ourselves alone having an inside and nature having none. Schwaller de Lubicz tells us that with the 'intelligence of the heart' we can 'tumble with the rock', 'seek light with the rosebud' and 'expand with the ripening fruit'; that is, with the 'intelligence of the heart', we can *get inside things*. Barfield is saying that what the history of language has told him is that, at an earlier time, we all shared in this intelligence. What

Müller and other language theorists saw as metaphor was this ability to 'see' nature's inside. And not only to see it, but to share in it. Folk tales and myths about nature spirits, fairies, elementals, and legends of the gods walking the earth, are all echoes of a time when we and the world shared a continuity of consciousness, what Barfield called 'original participation'.

Representation

To this idea of an evolution of consciousness, Barfield wedded what he calls 'representation'; I have written about this elsewhere.[22] In *Saving the Appearances* Barfield asks us to consider a rainbow. We know that what we see as a rainbow is the result of rain drops, sunlight, and our own vision. Rainbows we know are not illusions, yet we also know that they are not 'real' in quite the same way as other things are – say, a stone. Rainbows, unlike stones, appear only under special conditions. To see one, we need a sun shower and our line of vision to be at the right angle to it. Nevertheless, we all agree that rainbows are real. They are 'really there' and are not an illusion or hallucination. Now Barfield asks us to consider a stone, or, for that matter, anything that we all agree 'really exists' and which we 'really' see. We know what a rainbow is 'made' of. What about a stone?

Even though to our touch a stone is a hard, solid object, science tells us that it is really made of molecules, which are themselves made of atoms, which are themselves made of electrons, protons, and neutrons, which are themselves made of even smaller elementary particles, which are themselves held together by the Higgs boson which, we're told, is a kind of glue, giving other particles their mass. And there may indeed be even more elementary particles, or, if string theory is correct, oscillating strings or 'superstrings'. These particles or strings, however, are not solid things that are merely infinitely small. They are 'probability waves' 'quantum fluctuations', charges of energy, dimensionless points, or other kinds of things which are not really things at all, and are best described using arcane mathematical equations. The point is that, although when I reach out and touch the stone it feels solid and has a certain weight, what is 'really real' about the stone, according to science, is this fundamental non-physical 'something' of which it is made. The solid, heavy stone that I pick up is a 'representation' of this 'something', provided by my consciousness.

Years ago, in *The Nature of the Physical World* (1928), the scientist Sir Arthur Eddington explained that the desk he was sitting at was really two desks. One was the solid object on which his paper and pencil were resting, the other was made of atoms which, at their level, existed as stars do in space, with huge gaps of emptiness between them. Paradoxically, Eddington's table was made more of empty space than anything solid. Yet his hand and pencil did not pass through it. That is because his consciousness very obligingly created the solid table for him, just as it very obligingly created his solid hand, and the solid pencil it was holding (they, too, were made of atoms). Or, more precisely, it 'represented' the non-solid 'something' of which the table is made as solid. It doesn't make the 'something' but forms a picture of it. Just as our consciousness forms a rainbow out of rain, sunlight and our angle of sight, it also forms Eddington's table, and everything else around us, out of the particles of which they are made. What we take as nature, the solid, physical, material, phenomenal world around us, is, science tells us, a representation of this fundamental 'something'.[23] Again, consciousness does not simply mirror a world already made, but reaches out and forms a world out of what is 'there'. As the phenomenologist Paul Ricoeur remarked of Husserl's 'intentionality', 'seeing itself is discovered as a doing, as a producing – once Husserl even says "as a creating". Husserl would be understood ... if the intentionality which culminates in seeing were recognised to be a creative vision.'[24] In one sense then, the entire observable universe around us is a kind of rainbow. Or, as Eddington said: 'The stuff of the world is mind-stuff.'

A world without consciousness

Barfield points out that although scientists agree that this solid world we encounter is really only a picture our minds provide of a world that is fundamentally nothing like it – something they have been telling us for decades – when talking about something like consciousness, they seem to forget this. For one thing, they tell us what the world was like before there was any consciousness like ours around to represent it. But their accounts of this world – the prehistoric world of dinosaurs or even further back before life appeared – are, in all essentials, very much like that of a world with a consciousness like ours around to represent it. They forget that the world we see is a representation that our consciousness makes, and assume that such a world is more or less the same as a world without our consciousness. But we have no idea what a world without our consciousness to represent it would be like. (And even if we could go back to the 'dawn of time' to see how things were, it would still be our consciousness checking things out.) And, as we've seen, when talking about the origin of consciousness, they say it emerged somehow out of matter – in this case the brain – forgetting that the neurons and everything else we say the brain is made of are a representation of the non-physical 'something' that is the bottom line 'really real' thing. But if the hard, solid, physical world is a representation our consciousness provides of something that is fundamentally not hard or solid or anything sensuous, how can it be the origin of that consciousness? The physical world cannot be the cause of the consciousness that creates it. Logic alone tells us that. But scientists and philosophers like Francis Crick and John Searle seem to have forgotten this. People like Blake, we know, are aware of it, but then, Blake was only a poet.

As mentioned, scientists have been aware of the participatory relationship between the world and our consciousness, at least since Werner Heisenberg hit upon his 'uncertainty principle'. Heisenberg's principle undermined the nineteenth century notion of the scientist as some detached, objective, recording machine, whose subjectivity has been dutifully subtracted from his observation. Heisenberg recognised that the observer and the observed are in an inescapable intimate relationship, at least at the quantum level. He asks us to imagine a microscope able to detect electrons. Assume you want to observe an electron 'in the wild', as it were, as if you were David Attenborough and were trying to sneak up on it. The light necessary to see the electron travels down the microscope. But since the light itself is made of particles – photons – that have a certain energy, it collides with the electron and interferes with it – or, in Barfield's term, participates with it. Hence your observation of the electron alone has altered it. There is no such thing as a detached observer, at least when it comes to electrons.

A participatory universe

Some scientists have taken this realisation and built on it considerably. We can even say that the theoretical physicist John Archibald Wheeler (1911–2008) borrowed Barfield's term 'participation' and ran with it, bringing it into areas of speculation that Barfield himself may have stayed shy of.[25] Wheeler is responsible for coining such terms as 'black hole', 'quantum foam', and 'wormhole', but what concerns us here is his notion of a Participatory Anthropic Principle, or PAP for short, an acronym opponents of his views may find consoling. The 'anthropic cosmological principle', we know, argues that our universe is so arranged that intelligent life – such as ourselves – had to arise in it. Wheeler takes this further and says that our universe is the way it is *because* of human consciousness, a notion that, if expressed by a spiritual, metaphysical, or 'occult' thinker would be laughed at, but which, coming from a scientist, seems acceptable, or at least worthy of consideration. Wheeler speaks of a 'participatory universe' and tells us that the term 'spectator' should be stricken from the scientific record and 'participator' put in its place. Wheeler proposed that not only does our consciousness participate in bringing into being the world we experience now – in what is known in particle physics as the 'collapse of the wave function', it is the interaction of the observer with the observed that decides whether or not, say, a photon will behave as a wave or as a particle (Bohr's 'complementarity'.) It is somehow responsible for bringing the *past* into being too.[26]

In his fascinating book *The Labyrinth of Time: The Illusion of Past, Present and Future* (2012) the writer Anthony Peake describes an experiment which supports Wheeler's view. In 2007, at the École Normale Supérieure de Cachan in France, a team of scientists shot photons through a device known as a 'beam splitter', a kind of 'half-silvered' mirror. This allows half of the photons to pass through unaffected, while the other half is reflected off it. Which photon is reflected and which passes through is unpredictable, and large numbers are needed to get the 50/50 effect. When unobserved, the photons are in a wave state, and hence produce 'normal' wave behaviour, such as creating interference patterns when mingling with other electromagnetic waves. As mentioned above, it is only when an observer is added that the 'wave function' collapses, and the photons appear to act as particles. After hitting the beam splitter, both the photons that pass through and those that are reflected are directed to a series of prisms and mirrors that lead them to a second beam splitter some 48 metres away. Here the 'split' photons are reunited, and the usual interference patterns are detected. In the experiment, this second beam splitter could be turned on or off randomly, *after* a photon had already left the first splitter. If the second splitter was left on, the two paths taken by the photons were brought together and sent to a single detector, where they created interference patterns – that is, acted like waves. If the second splitter was off, the two paths were kept separate, and the photons were sent to different detectors, where they acted like particles. The second splitter 'decided' which path the photons would take and whether they would behave like waves or particles. But the photons had already started on their journey when the second splitter was turned on or off. The *future* act of turning the splitter on or off 'decided' the *past* action of the photon's path.

As Peake shows, Wheeler proposed a 'thought experiment' to support his 'participatory' view, involving the results of what is known as the 'two slit' experiment, but on a cosmic level.[27] In the 'two slit' experiment, photons are shot at a barrier in which two slits have been cut, forcing the photons to pass through them. If two particle detectors are placed on the other side of the barrier, a photon passes through one or the other of the slits, and is detected by one or the other of the detectors. But if a wave detector – a screen – is used instead of the particle detectors, the photon then passes through *both* slits and leaves an interference pattern on the screen. In other words, either the photons 'know' whether a screen or a detector is waiting for them, and act accordingly, which seems unlikely. Or the act of placing either the screen or the detectors 'makes' the photons behave either like a wave or a particle. The experimenter's decision determines the results.[28]

Wheeler took this and applied it to light coming from a immensely distant source, quasars, quasi-stellar objects that are very far from us but emit tremendous amounts of electromagnetic energy. Quasars are so distant from us that the light they emit and which we can detect takes billions of years to reach us. Wheeler believed that whether photons came from a present light source or from quasars, they would behave in exactly the same way as in the 'two slit' experiment. Yet the photons from the quasar started their journey to earth billions of years before life appeared on our planet. Just as the observer in the 'two slit' or 'beam splitter' experiments 'decides' whether a photon acts as a wave or a particle, so too would the observers in Wheeler's thought experiment. But their observation is now 'deciding' on the character of light that left its source billions of years before Wheeler or anybody else existed. For Wheeler this led to the conviction that 'we are participators in bringing into being not only the near and here but also the far away and long ago. We are in this sense, participators in bringing about something of the universe in the distant past ...'[29]

Final participation

Barfield would no doubt recognise the significance of Wheeler's speculation, but his own concern was more about the phenomenal character of our experience – the macro world of nature – than the quantum, and he was more concerned with the future than the past. What concerned Barfield was our 'representation' of that strange 'something' that is not ourselves, what he refers to as the 'unrepresented'. If, Barfield reflected, our picture of the phenomenal world, the world of clouds, trees, stones, and also things like quasars, is a product of our consciousness representing the 'unrepresented'; and if, as Barfield believed, consciousness evolves, then our current representations were correlative to our current consciousness. Which meant that any future representations were dependent on our future consciousness.

Barfield believed that the trajectory of the evolution of consciousness was shaped like a U, and consisted of a movement from above to below and then back up again on the other side. He believed, as did Max Scheler, Jean Gebser, Rudolf Steiner, and other thinkers I've discussed in this book, that consciousness began in a unified state with what we call 'nature', and then gradually separated from this, until, with ourselves, we have reached the bottom of the U, when we are more or less completely separated from it – an extreme example of this being the 'nausea' depicted in Sartre's novel. It is not the case, however, that we no longer participate with the 'other'. We do, but we are unaware of our participation, we are not conscious of it. Earlier we were unaware of it in a conscious sense as well, but we *felt* it, just as animals are unaware of their participation but are nevertheless 'in' it (they always say 'Yes' to reality), or as children are unaware of their ouroboric relation to the world, but nonetheless feel it. As we are now, we neither feel it nor are aware of it. Hence our alienation. But, as we've seen throughout this book, there are moments when we do become aware of it – even if we do not speak of it specifically as such. And poets, philosophers, artists, and mystics have felt it too. Barfield believed that, having hit rock bottom, we were slowly working our way up the other side of the U, toward what he called 'final participation', a kind of consciousness in which we are consciously aware of and in control of the participation we had hitherto only shared in unconsciously. We are, we can say, slowly making our way to the regained 'grace' that Heinrich von Kleist spoke of. And, as Kleist said, this was something like the consciousness of a god.

With this, however, comes responsibility, an important obligation for any caretaker of the cosmos. Once we become aware of our role in representing the world, a responsibility is incumbent upon us, a responsibility of the imagination.

We have to remember that Barfield isn't saying that we 'create reality', as some superficial new age teachings suggest, or as the equally superficial frivolity of postmodernism argues. I cannot, now, through an act of will, turn my studio into a castle or my sofa into a rocket ship. Such an 'anything goes' sensibility is the exact opposite of the responsibility of the imagination I am speaking of. Barfield summed up his concern in an elegant statement the sobriety of which belies its importance. 'If the appearances are,' he wrote, 'correlative to human consciousness, and if human consciousness does not remain unchanged but evolves, then the future of the appearances, that is, of nature herself must indeed depend on the directions which that evolution takes.'[30]

The fate of the Earth

This statement warrants repeated re-readings. If our consciousness is responsible for the phenomenal world, for stones as much as for rainbows, and if, as Barfield has shown, consciousness has evolved to its present state, with the understanding that it is evolving still, then it follows that the fate of the phenomenal world, of nature and the cosmos, is not only in our hands, but in our *minds*. If this is the case, then we need to be 'environmentally correct', not only through all the 'outer' ways we are all familiar with: saving the rain forests, conserving natural resources, avoiding waste, developing renewable energy, and all the other important and essential efforts through which we can fulfil what Rudolf Steiner called 'our responsibility to the earth'. These are of course needed and there is no call to avoid them. But we need to think of the 'inner' ways we can 'save the planet' too. This is because, if before we participated with the world unconsciously, now, having separated from it in order to develop our independent consciousness, we are entrusted with the task of learning how to participate with it *consciously*. Or, in the terms I used earlier in this book, to perform *tikkun*. This means making our own consciousness more alive and active, more intentional and purposive, which means taking responsibility for it. I can no longer rest in the 'natural standpoint' and accept as simply given the world I see, just as I can no longer see my consciousness as a passive mirror, simply reflecting that world. If the world is given, it is given *by my consciousness*, by that part of my mind that is involved in 'representing', and it is only my ignorance of this fact that allows me to 'accept' the world without a moment's thought. But anyone who has followed this book so far can no longer claim such ignorance. If anything I have written here has any value, we have gained some knowledge, and with that knowledge come consequences. It presents a good that we know, and as we know it, we should then do it.

One of the most bizarre things Rudolf Steiner said was that the future physical body of the earth would be determined by the kinds of thoughts people think now, just as the present physical earth was determined by the thoughts of earlier people. (Rudolf Steiner also said that our present Earth was 'dying', in preparation for the next stage in its evolution.) We can take this with several grains of salt or chalk it up to madness. But if it is true, as Barfield, Blake, Gebser, Wilson, and others have argued, that our consciousness is intimately involved with what we call the physical world, in ways we are only just beginning to grasp – and as I have shown, science itself is to some extent on our side here – then I think it is a reflection we would ignore at our peril.[31] It is also a reflection that argues very strongly against the idea that nature or the earth would somehow be better off without us. On the contrary, it is considerable support for the idea that we are essential to both, that we are, in fact, entrusted with something more than the responsibility of caretakers, but of co-creators. Earlier we saw that Julian Huxley and Abraham Maslow believed that humanity had reached a stage at which it had become responsible for its own evolution. Barfield, Steiner, Gebser, Wilson, and others agree. But to this we should add that we have become responsible for the evolution of the cosmos too. As Barfield wrote: 'The future of the phenomenal world can no longer be regarded as entirely independent of man's volition'.[32]

This is about as far from the standard materialist scientific position as we can get. But if Barfield is right, then entertaining ideas about ourselves in which we are only 'genetic drift' or a chance outcome in a cosmic 'Monte Carlo game', or of our consciousness as merely 'the behaviour of nerve cells' or something like bile, will clearly not do. If we think of our own minds – the seat of 'representation' – as bile, or any other kind of 'epiphenomenon', in what way will we think of anything else? If we diminish the importance of that which is responsible for representing reality, what will our representations of it be like? To in any way applaud or revel in our insignificance, finding in that some virtue, or to see ourselves as fleeting nothings in a meaningless universe that 'just happened', is to embrace Maslow's Jonah Complex on as fundamental level as we can imagine. It would be an expression of the kind of false humility Saint-Martin warned of, which breeds the laziness and cowardice that allow us to avoid the responsibilities and duties incumbent upon ourselves as repairers of reality. It would, in fact, be the equivalent of leaving the cosmos in the lurch.

Completing

Readers may find the above reflections about participation somewhat extreme, yet there are other ways in which we can recognise how our consciousness and the world participate with each other. One of the clearest and most effective ways of experiencing this participation is those strange 'meaningful coincidences' that C.G. Jung dubbed 'synchronicities', and which I mentioned earlier. Anyone who has ever had a thought in mind and then seen it in the outer world at the same time, with no causal link between the two, knows what I'm speaking of. On a more existentially immediate level than the quantum, these strange episodes are convincing proof – at least for myself – that there is some continuity between what is going on 'in my head' and what is happening 'out there', that the wall separating the two is, at least at times, permeable. Indeed quantum physics and other scientific ideas have been offered as aids in explaining or accounting for this unsettling experience. Jung's book, *Synchronicty: An Acausal Connecting Principle* (1952) does precisely this, but to my mind it is ultimately unconvincing. Not that I am unconvinced about synchronicities – hardly. I have collected notes about my own for decades now and am as convinced of their reality as I am of anything else. But I don't think bringing in the strange behaviour of particle physics or postulating some kind of universal 'acausal connecting principle' helps much in understanding them, although these ideas can help us understand that there is less of a distance between the inner and outer worlds than we believe. I'm more inclined to accept the idea that synchronicities are little taps on the head or pats on the back from some guardian angel, letting us know that, although things seem confused and muddled now, we are on the right path after all. At least I have discovered that this is a more fruitful way to look at them, rather than as merely anomalous phenomena calling out for an explanation.

Synchronicities are more than mere coincidences, however surprising; they are *meaningful* coincidences, and that meaning is usually of a very intimate, personal kind, the kind, Jung discovered, that can make all the difference to someone in the grip of a crisis, or looking for some hint of what path to take with their lives. This means that they are not merely 'weird stuff' but have a deeper, more existential import. This suggests to me that in some way, my own mind is involved in making them happen, that they are not simply something that happens *to* me, but that on some level, my own mind reaches out and 'arranges' events in order to get the message across. How this can take place, the necessities and logical conundrums involved, are indeed baffling, and to be honest, I do not have a good argument for it. But perhaps this is one of the purposes of synchronicities, to punch holes in our logic – an essential tool, but sometimes limiting – and to allow *something else* to arise. Colin Wilson suggests that when we are optimistic, living creatively, and are filled with a sense of purpose – more 'fully human' – synchronicities happen more, and I have found this to be the case. Most of the ones I've recorded in recent years have happened while I was working on a book; appropriately I noted a few in my book on Jung.[33] In a way we can see this, as Wilson does, as an extension of Husserl's intentionality, as a more direct and unmistakable way in which the mind 'reaches out' and shapes reality.

But even in our more everyday experience we can see how our mind *adds* to reality, how it, as it were, 'fleshes it out'. In his book *Beyond the Occult* (1988) Wilson speaks of what he calls 'completing'. 'Our minds,' Wilson writes, 'are inclined to accept the present moment as it is, without question.' This is what Husserl calls the 'natural standpoint'. 'In what we might call "neutral consciousness"', Wilson says, 'we accept the present moment *as if it were complete in itself*.' [italics in original.] But, Wilson tells us, 'a little reflection reveals that this is a mistake of gargantuan proportions'.[34] The truth, Wilson says, is that the present moment is always incomplete, and that 'the most basic activity of my mind is "completing" it.' Wilson asks us to imagine an alien somehow suddenly finding himself on a bus in London. The world, which we take for granted as given and complete in itself would appear to him as chaos. He would have no idea of what things were – the same things that we believe just appear as full and completed to our consciousness. If he saw someone walking a dog, he would have no idea that the dog was a pet, and that it wasn't walking the man. If he saw someone enter a car, he wouldn't know that it wasn't going to eat him. If he saw someone speaking into a mobile phone, he wouldn't know that it wasn't conscious, and answering him back. Our alien would know none of the things we do and which we assume are just there, reflected by our consciousness. But it is our *knowledge* of things that gives them their meaning, their depth and reality, and provides the connections linking them to everything else. (We can say that Sartre's 'nausea' is a result of faulty 'completing'.) Try making the experiment of looking at something and subtracting from it everything that you *know* about it. Try, that is, to see it just as 'given'. It is very difficult and is the beginning of phenomenology.

We don't even need to invoke an alien. A baby is equally unaware of the meanings of things that we take as given. This is because a baby's consciousness and inner world is not yet developed enough to perform the acts of 'completing' that we engage in all the time. That is precisely what 'growing up' and 'getting an education' is: developing our inner repertoire of meanings with which we can complete the world around us. A tourist in London who knew little of its history may enjoy a stroll along the Thames, but someone who had some background in London's rich past would see more, would grasp the reality of Shakespeare's Globe Theatre, as he would that of the Tower of London, or the old docksides to the east. As Rudolf Steiner said in *Goethe's Theory of Knowledge,* 'When one who has a rich mental life sees a thousand things which are nothing to the mentally poor, this shows as clearly as sunlight that the content of reality is only the reflection of the content of our minds'. Philosophers and scientists who want to reduce our minds to the content of reality – the physical world – should re-read this sentence several times. We do not reflect a reality that is already there. We provide that reality. Again, as Steiner said, 'We receive from reality only the empty forms'. The 'empty forms' are the 'bare facts' of things, before our minds complete them. Coleridge said the same thing when he wrote in 'Dejection, an Ode' (1802), unaware that he was echoing Blake (whose work he was ambivalent toward): 'We receive but what we give/And in our life alone does Nature live'.[35]

Making things interesting

Our ability to complete reality, however, varies with our own vitality and willingness to make the effort of imagination required, as well as with our attitude toward the world. If I am tired or distracted I can read a paragraph several times and still not 'get it'. If I believe whatever I am looking at is uninteresting, then I will certainly make very little effort in contemplating it. At other times, my mind may be so active and vital, and I may be so convinced of something's significance, that I see so many connections between things that I am overwhelmed. This is what Ouspensky felt when contemplating his ash tray, and what we feel to a lesser degree when we are optimistic and 'full of life' and find everything around us 'interesting'. We are completing things a bit more than usual, and sense the hidden meanings hovering around them. That is what their 'interestingness' *is*. We find something interesting because we are vaguely aware that, in the old saying, there is more to it than meets the eye, an adage we should make into a mantra, and an example of language itself coming to our aid. This is also what Whitehead meant when he said that we cannot arrive at 'an adequate description of finite fact' because such a description relies on a 'background of presupposition which deifies analysis by reason of its infinitude' – in other words, its 'interdependent connection' with everything else. To complete something completely would be to spell out its connections to everything else in the universe. It would, in essence, mean coming up with a theory of everything, which could start with anything at all.

Many people only see what is right in front of them; their imagination and will are too feeble, or they are too lazy, to see more than what is just 'there'. But imaginative, active people are always seeing the meanings that radiate out of the given, and are always connecting them to the rest of reality, either in the past, the present, and the future. People of this sort can see the consequences of their actions and can recognise more in things than their duller fellows can. On Primrose Hill, not far from where I live in Hampstead, there is a quote from Blake carved into a stone curb overlooking the view of London's growing skyline. It reads: 'I have conversed with the spiritual sun – I saw him on Primrose Hill'. Many people taking in the view read the quotation and move on. Some wonder what it means. Some don't notice it at all. But if, like myself, they are readers of Blake and are aware of his mythology and ideas about perception, and know that he had his experience with the 'spiritual sun' two centuries earlier, right where they are standing, they undoubtedly see more. They can see Blake standing there, looking out over a London that was very different from the one they see, and seeing a sun that was also very different from the one they see. And they see more, precisely in the way that Blake saw the spiritual sun: with their imagination, that is, with their mind alive and active and doing its job of completing the world. Bernard Shaw once quipped that a picture gallery is a dull place to a blind man. Without making the effort of completing, we are all in that department. We feel dull and bored and make the mistake of assuming that the world we are confronted with is dull and boring 'for real'. It isn't. By not completing it with our imagination we allow ninety per cent of reality to lie fallow. Again, Goethe said it long ago. When a bored, exhausted Faust confronts a world he thought he knew, and is contemplating ending it all, he is told: 'The spirit world shuts not its gates. Your mind is closed, your heart is dead.'[36] When the mind is open and the heart alive, then the gates to the spirit world – the doors of perception – are flung wide, and everything is seen as it is, infinite.

New dimensions

For some philosophers, the act of completing – although they do not use the term – does its work by adding a new dimension to reality. At least this was what Ernst Cassirer believed. Like Whitehead and Bergson, Cassirer is another important thinker of the twentieth century who is little read today. His most popular book, *An Essay on Man* (1944), is a work of 'philosophical anthropology' like Berdyaev's *The Meaning of the Creative Act* and Scheler's *Man's Place in Nature* (1928). In it Cassirer argued that: 'as compared with the other animals, man lives not merely in a broader reality; he lives, so to speak, in a new *dimension* of reality.'[37] This was, Cassirer argued, a symbolic dimension, that is, a mental, spiritual dimension. Man, he said, is a symbolic animal. Whereas other animals relate to reality through instinct and direct sensory perception, man understands himself and the world through creating a symbolic universe, a universe of knowledge, meanings and ideas, something Cassirer spelled out in detail in his *magnum opus, The Philosophy of Symbolic Forms* (1923–1929). Language, science, mathematics, mythology, art, philosophy, religion, literature: culture in all its forms constitutes the symbolic world in which we live and which is unavailable to other animals.

At the beginning of *An Essay on Man* Cassirer echoes an idea mentioned earlier in this book: that mythological accounts of the origin of the world are invariably linked to an account of the origin of man. 'In the first mythological explanations of the universe', Cassirer writes, 'we always find a primitive anthropology side by side with a primitive cosmology. The question of the origin of the world is inextricably interwoven with the question of the origin of man.'[38] In the mythological mind, Cassirer discovered, the world and man are one, something with which Barfield, Gebser, Scheler, Blake, Swedenborg and other thinkers discussed in this book would agree.

This is once again a recognition that without man there is no world. It is also an echo of Saint-Martin's dictum that we should not explain man in terms of the world, but the world in terms of man. Cassirer shares with Barfield, Scheler, Gebser, and Rudolf Steiner the belief that we began in a kind of ouroboric embrace with nature and only gradually separated out from it to carve out 'the world'. 'We cannot discover the nature of man in the same way that we can detect the nature of physical things,' Cassirer writes. 'Physical things may be described and defined in terms of their objective properties, but man can be described and defined only in terms of his consciousness.'[39] Like Berdyaev, Scheler, and others, Cassirer knows that the task of creating a 'philosophical anthropology' – a true account of man – is not a mere 'intellectual' effort. It is a 'clash of spiritual powers' within which 'the whole destiny of man is at stake'.[40] To which I would only add, echoing Barfield, that not only the destiny of man is at stake, but the destiny of the cosmos itself. 'If we were to know all the laws of nature,' Cassirer writes, 'if we could apply to man all our statistical, economic, sociological rules, still this would not help us to "see" man'.[41]

Yet, as has been pointed out in this book, that is precisely what the explainers of man and consciousness want to do and what they persist in declaring to be the only way in which to see man. But to see man in this way is to not see him at all. Man cannot be explained in terms of the laws of the universe, because that universe and its laws are themselves the work of man. It is through man's consciousness, through his power to complete the world that a universe comes into existence. There is no universe for animals, nor for planets, stars, nor the billions of galaxies whose immense size and unthinkable distance from us seem to dwarf our own humble reality. But they dwarf us only if we ignore that other half of our being, that new dimension that is possible only through the human mind, an insight that Cassirer shared with Julian Huxley and H.G. Wells. As we saw, Huxley believed that with the appearance of man, something different had entered the evolutionary process. And Wells once remarked that just as a bird is a creature of the air, and a fish a creature of the sea, *man is a creature of the mind*. We share the air and the sea with the birds and the fish, but they cannot share the world of mind with us. It is a completely new dimension, a new world of reality, accessible only to us or to other 'creatures of the mind'. This is something so remarkable that to want to deny it seems incredible. Without us reality would be less. Perhaps this is why, in creating the world, God purposefully made a mistake, so that, as some Kabbalists believe, we would be called in to finish the job. Reality needs us. To try to explain ourselves in terms of what we share with the birds and the fish – as well as the worms of the earth – may indeed tell us much, but it will not touch the *essence* of being human, or at least of being 'fully human'. This is in fact the key message of this book. I do not deny that we share much with other animals. But Cassirer, Scheler, Berdyaev, and other thinkers I have drawn on tell us that what makes us 'us' is *not* what we share with animals, but what is unique to ourselves. And it is this uniqueness that people like John Gray, Jacques Monod, Stephen Hawking, John Searle, and other explainers deny.

'Not the material but the human world,' Cassirer tells us, 'is the clue to the correct interpretation of the cosmic order'.[42] Cassirer explains that this is because speech, language, is at the heart of such an order, something we have seen in Barfield, Heidegger, Saint-Martin, and Pythagoras, who started it all. Language is so fundamental that we cannot even begin to grasp its importance. 'We must ... understand what speech means in order to understand the "meaning" of the universe,' Cassirer writes. If we fail in this, if we do not find the way to this meaning through language and ourselves, Cassirer warns, then we will 'miss the gateway to philosophy'.[43] Miss, that is, the gateway to wisdom. The central importance of speech is that it is a creative act, and we are the only ones who can make it. It is Adam who gives names to his fellow creatures in the garden, not God. In this sense, the meaning of the universe depends on us; we will not find it in the natural world, nor in the stars bordering the rim of the observable universe. Only we can ask the question 'what is the meaning of the universe' and only we can provide an answer. Speech is an expression of man giving shape and order to the world, something, Cassirer suggests, that also comes through in art. 'Like all the other symbolic forms,' Cassirer writes, 'art is not the mere reproduction of a ready-made, given reality. It is one of the ways leading to an objective view of things and of human life. It is not an imitation but a discovery of reality'.[44] 'The artistic eye is not a passive eye that receives and registers the impression of things,' he tells us. 'It is a constructive eye, and it is only by constructive acts that we can discover the beauty of natural things.' This remark leads us back to our discussion of how consciousness itself is responsible for nature and indeed, for the entire cosmos we are endeavouring to save.

Our place in the universe

We can save the universe, we can repair it, take care of it, redeem it and awaken it from its trance by becoming aware of our creative contribution to reality and by intensifying our consciousness to such a degree that we never lose sight of this fact. That is our place in the universe, and that, I submit, is the kernel of the work of the thinkers I have discussed. It is not so much that there is something particular and specific that we must do to take care of the cosmos. As Gershom Scholem says, there is no specific act of *tikkun*. We can carry on doing what we normally do, but we must do it more consciously. We must become aware of ourselves as active minds completing the world which, without us, would be empty, mute, and, from this perspective, an immense waste of time, nearly 14 billion years by most accounts.

I would even go so far as to suggest that the only reason the universe is mysterious or even interesting is because we are in it. Would Big Bangs, black holes, quasars, and Great Walls mean anything if they didn't present us with the question of how we fit into this picture and why we are here? Without us the universe is merely an immense flux of cosmic gas and dust, going nowhere, doing nothing, and to be impressed with that is a delusion, is, in fact, idolatry, the kind Blake attacked in his 'mental fight'. Or, if it is not this, then merely a collection of mathematical equations that need a consciousness like our own to decipher them. To not be a waste, to not be a failure, it needs us. Each act of creative consciousness is an act of *tikkun*, an act of redeeming the universe from pointlessness, and we can perform it at any time. We do not need John Wheeler's speculations about how we can change the past. We can change it each time we become fully human, by making everything that led up to that moment worthwhile, transforming meaningless events into a chain of destiny. That is the alchemical secret of turning lead into gold, or water into wine, or death into life. We cannot shirk this duty. The universe needs our help and we cannot let it down. In the last chapter of this book I will take a look at what might be in store for the universe, and ourselves, if we don't get to work.

7. As Far as Thought Can Reach

In the first decades of the last century, the question 'What is Man?' took on a new and confusing complexity. Developments in science, in psychology, in anthropology, sociology, history and other related fields had increased at such a speed and had produced such a wealth of new and complicated data that the idea of arriving at a complete idea of man, which could accommodate all this new information, seemed beyond our grasp. The earlier, simpler definitions of man as the rational animal, or as a creature of God, or as the bearer of certain inalienable rights, could no longer suffice, and it seemed that there was no single answer to the question of who or what we are. There was economic man as seen by Marx, man the pursuer of the will-to-power as seen by Nietzsche, man the 'trousered ape' as seen by Darwin, man the repressed neurotic as seen by Freud, and many other versions. But there was no agreement among them, and although each individual perspective seemed to provide some stable foundation for understanding ourselves, one had only to turn to the next to find the first undermined. If the admonition to 'know thyself', the bedrock of western philosophy, had always been difficult to fulfil, by the early part of the twentieth century, it seemed downright impossible. It was in this climate of self uncertainty that philosophers like Cassirer, Scheler, Berdyaev, and others sought to anchor our identity in what they called a 'philosophical anthropology'.

The irony was that it was knowledge itself that had obscured our self-knowing. In a chapter of *An Essay on Man* entitled 'The Crisis in Man's Knowledge of Himself', Cassirer wrote that: 'No former age was ever in such a favourable position in regards to the sources of our knowledge of human nature. Psychology, ethnology, anthropology, and history have amassed an astoundingly rich and constantly increasing body of facts. Our technical instruments for observation and experimentation have been immensely improved, and our analyses have become sharper and more penetrating.' 'When compared with our own abundance,' Cassirer wrote, 'the past may seem very poor.' Yet Cassirer also pointed out that 'our wealth of facts is not necessarily a wealth of thoughts' and he warned that 'unless we succeed in finding a clue of Ariadne to lead us out of this labyrinth, we can have no real insight into the general character of human culture' – Ariadne's thread, of course, being the means by which the Greek hero Theseus was able to escape from the labyrinth of the monstrous Minotaur. Without such a thread, Cassirer believed, 'we shall remain lost in a mass of disconnected and disintegrated data which seem to lack all conceptual unity'.[1]

In saying this Cassirer was echoing a concern Max Scheler had voiced some years earlier. In *Man's Place in Nature* Scheler wrote that: 'In no other period of human knowledge has man become ever more problematical to himself than in our own days'. 'We have,' Scheler says, 'a scientific, a philosophical, and a theological anthropology that know nothing of each other. Therefore we no longer possess any clear and consistent idea of man. The ever-growing multiplicity of the particular sciences that are engaged in the study of man has much more confused and obscured than elucidated our concept of man.'[2]

In *Being and Time* (1927), Martin Heidegger, a colleague of Scheler and an opponent of Cassirer, devised his own philosophical anthropology, yet he believed much as they did. 'No other time,' Heidegger wrote, 'has had so much and varied knowledge of man as ours. No other time has expressed its knowledge of man in so impressive and striking a way. No other time could supply this knowledge so quickly and easily. But on the other hand no other time knew less what man is than ours. Man has never become so questionable as in our time.'[3]

This sense of crisis, of confusion and ambiguity about man and our place in the world, permeated much of the thought of the first half of the last century. Whitehead wrote about it in *Science and the Modern World* and even Husserl, who sought to secure for philosophy an unshakeable foundation of certainty, succumbed to the general doubt, as can be seen in his late unfinished work *The Crisis of the European Sciences* (1936). If Pico della Mirandola could declare, in his *Oration on the Dignity of Man*, that what was most worthy of wonder was man, by the time Cassirer raised his concerns, one could only wonder what man was.

Cassirer believed that this mass of disconnected information about ourselves led to a complete 'anarchy of thought'. The 'modern theory of man has no intellectual centre', no 'general orientation' and left one with a sense of conceptual paralysis.[4] But such a dizzying explosion of unrelated facts, Cassirer believed, posed more than a philosophical or scientific problem. It also posed a threat to our ethical and cultural life. If we do not know who we are, how can we lead ethical lives? How can we continue to create, to inhabit the symbolic universe that is uniquely ours? What values can we uphold and strive to actualise? If we do not know what the human is, how can we recognise the inhuman? When Cassirer's *Essay* appeared, he had left his native Germany and had settled in America, in order to escape a brutal regime that seemed to have taken advantage of our uncertainty about who man is, in order to actualise its own inhuman version of him. Hitler's Germany wasn't the only place where such a redefining of the human was taking place. In Stalin's Russia what it meant to be man went through some changes too, and in many parts of the world, the process is still going on.

The last Men

I do not think it requires a great deal of argument to show that today, after nearly a century of living through this identity crisis, we are nowhere near resolving it. Which is not to say that we are still troubled by it. If anything, the mass of disconnected information about ourselves and practically everything else has increased a hundredfold and we are subject to new revelations about who we are – or aren't – almost daily. What Cassirer or Scheler would have made of the internet, Wikipedia, Google, and other sources of our increasing 'wealth of facts' is unknown, but I suspect these developments would only have motivated them to increase their efforts to warn us of the dangers of drowning in a flood of information that does nothing to help us understand ourselves but rather only adds to our already weighty burden of confusion. With each new 'discovery' binding us ever more closely to our selfish genes, to our animal past, to the molecules in our brain's nerve cells, or to an empty, meaningless universe the secret of which can be found in an infinitesimal particle – or perhaps in the vibrations of strings – the idea of 'knowing ourselves' in the Socratic sense, seems increasingly out-dated, much as the idea of sin already is in our secular world. Cassirer believed that man 'is that creature who is constantly in search of himself – a creature who in every moment of his existence must examine and scrutinise the conditions of his existence.'[5] Yet today, in our post-modern world, it seems we have already discovered who we are, if the 'scientific' picture of ourselves that enjoys increasing authority is anything to go by. And the answer is: nothing special. This redefinition of man will, I think, be involved in some radically different forms of human life as this century moves on.

This conclusion may relieve us of the burden of mystery that troubled our ancestors, and the dread and anxiety that distressed existentialists only half a century ago may no longer be ours. Much of post-modern thought sees the kind of uncertainty about ourselves that kept angst-ridden earlier generations up at night as little more than a cosmic joke, and wonders, with much of modern cosmology, what the fuss was all about. We accept that there is no reason for our existence and that the universe is too big to even notice we are here, if it could notice anything at all. There's no reason for anything, it tells us, so get over it and have fun. Post-modern thought has set us free. We have been liberated from the demand to be great and relieved of the obligation to be something better than ourselves. We can embrace our mediocrity with a good conscience and sleep better at night. What more can we ask of philosophy?

Nikolai Berdyaev had a word for this self-satisfaction. In *The Destiny of Man* (1937) he speaks of what he calls 'smugness'. We know of course what smugness is, but with Berdyaev it takes on a broader, more metaphysical meaning. 'Smugness,' Berdyaev writes, 'frees man from fear not through rising to higher spheres but through sinking to the lower.'[6] It is a way of acclimatising ourselves to a lower place, a kind of metaphysical sour grapes, a contentment with the banality of life, with Heidegger's 'triviality of everydayness'. It allows us to be at home in a world without meaning, without significance, without cosmic drama. It finds a cosy place for itself in the commonplace, where it can forget the existence of any kind of 'other', more important world, and feel thoroughly satisfied that it has done so. We may have emancipated ourselves from the existential cares of an earlier generation, but such emancipation 'brings with it the danger of vulgarising life, of making it flat, shallow and commonplace'.[7] Such, I believe, is our post-modern world. Post-modernism 'ironises' out all questions of meaning. It reduces everything to the 'been there, done that' mentality, and shrinks the world to a theory of everything that can fit on a T-shirt. It lets us off the hook. We no longer have to be good, just good enough. It lowers the existential bar, and moves the metaphysical goal posts closer, or gets rid of them entirely.

In *Thus Spoke Zarathustra* (1883–85) Nietzsche speaks of those he calls 'the Last Men' (some translations have the 'Ultimate Man'). '"What is love? What is creation? What is longing? What is a star?" Thus asks the last man, and he blinks. The earth has become small, and on it hops the last man, who makes everything small ... "We have invented happiness," say the last men, and they blink ...' 'No shepherd and one herd. Everybody wants the same and everybody is the same. Whoever feels different goes voluntarily into the madhouse.' 'One is clever and knows everything that has ever happened; so there is no end of derision. One still quarrels, but one is soon reconciled ...'[8]

Earlier Zarathustra had hoped to get the people interested in his idea of the *Übermensch*, the 'overman' who has transcended himself and become something *more*. But the people weren't interested. Instead, they wanted to know more about these 'last men' and they ask Zarathustra to turn them into them. The last men prefer comfort and pleasure to the hardships 'overmen' must pursue. Discouraged, Zarathustra leaves them.

Are we the last men? Do we prefer comfort and pleasure to facing up to the tasks of being 'fully human', which is really all that being 'overman' means? Ironically, today Nietzsche is seen as a forerunner of post-modernism, yet he has more in common with Blake, Kazantzakis, Berdyaev, Saint-Martin and others discussed in this book than with the thinking of the 'last men', who lack the will to be creators and avoid the challenge of giving meaning to their lives. Such irony isn't lacking in Nietzsche's career; we know he was appropriated by the Nazis, too. And he saw it coming. So perhaps it is appropriate that a prophet who warned us of the last men should be appropriated by the very last men he warned us about.

The end of things

That we live in a kind of 'end times', regardless of any actual imminent apocalypse, has been the case for some decades now. Thinkers as far apart as Michel Foucault and Francis Fukuyama have written about the 'end of man' and the 'end of history'.[9] Deconstructionism has declared the end of philosophy. Some critics have even written about the end of science.[10] Even bastions of western culture have accepted that things are in decline. In *From Dawn to Decadence* (2001), a cultural history of the west from the Renaissance to the present day, Jacques Barzun (1907–2012), witness to practically the whole of the last century, spoke of 'the Great Undoing', western culture's self-immolation at the hands of chat shows, gangsta rap, and post-modernism, and what he sees as an unravelling of five centuries of progress into a morass of conflicting values. In *Grammars of Creation* (2001) George Steiner argues that 'there is ... in the climate of spirit at the end of the twentieth century, a core tiredness'. 'We are latecomers,' he tells us, at the banquet of western thought, where 'the dishes are being cleared'.[11] Even pop music picked up the cue. In 1987 the band REM released 'It's the End of the World as We Know It (And I Feel Fine)', a post-punk stream of consciousness rap that seems to say everything's going up in flames but it doesn't matter. When the Sex Pistols released 'Anarchy in the UK' a decade earlier, the idea of social collapse was threatening. By now it's a dance tune.

But these are local events, related to the provincial affairs of western culture. What about the cosmos itself? The idea of the end of the world has been around practically since civilisation began, but since the nineteenth century its traditionally religious or mythical character has been superseded by a rigorously scientific one.[12] In 1865 the German physicist Rudolf Clausius introduced the notion of entropy. Clausius had observed that over time, in a closed system, organised energy – for example heat – tends to move toward a less organised, uniform state. (This is why a cup of coffee cools to room temperature.) As the 'second law of thermodynamics,' this suggested that eventually, the organised energy in the universe would dissipate until it formed a kind of lukewarm cosmic puddle, unable to support life. This irreversible process was known as the 'heat death' of the cosmos. For all the progress associated with the Industrial Revolution and the nineteenth century, a sense of futility had entered things, a feeling for which can be found in Matthew Arnold's famous poem 'Dover Beach,' with its images of 'ignorant armies clashing by night' upon 'a darkling plain'. The eventual outcome of the universe's 'heat death' was perhaps most eloquently, if depressingly, put by the philosopher Bertrand Russell. Because of it, Russell tells us, 'all the labour of the ages, all the devotion, all the inspiration, all the noonday brightness of human genius are destined to extinction in the vast death of the solar system, and that the whole temple of man's achievement must be buried beneath the debris of a universe in ruins – all these things ... are yet so nearly certain that no philosophy which rejects them can hope to stand'. Some years later, Russell made the point even clearer. 'The second law of thermodynamics,' he says, 'makes it scarcely possible to doubt that the universe is running down, and that ultimately, nothing of the slightest interest will be possible anywhere ... So far as scientific evidence shows, the universe has crawled by slow stages to a somewhat pitiful result on this earth and is going to crawl by still more pitiful stages to a condition of universal death'.[13] The universe may have started with a bang – although when Russell gave his lecture the notion of a Big Bang had yet to appear – but, to take a cue from T. S. Eliot, it will apparently end with a whimper.[14]

If this wasn't enough to suggest there was nothing much to get excited about, more recent ideas about the eventual fate of the universe seem to clinch it. One such is the notion of the Big Crunch. In this scenario, some billions of years from now, the universe will rebound on itself, like a rubber band stretched to its limit and then let go. The expansion of the universe started by the Big Bang will eventually slow down, and the gravitational pull of its matter will start the reverse process, with one possibility being that all the objects in the universe will collapse on themselves and create a single black hole – or 'singularity' – although what it will be a black hole *in* is unclear. Some scientists believe that this may then restart the cycle, with another bang and another universe, but by this time we, or anyone like us, will not be around to know. An even more recent contender, though, suggests that the Big Crunch will not be the way things go. It seems increasingly likely that, rather than come together in a great cosmic hug, all the interminable galaxies will move further away from each other, until eventually each one will be left all to itself, amidst vast black emptiness. This will be the result of what is called 'dark energy', another 'something' that science suspects is there, but knows very little about. Dark energy is a hypothetical form of energy that scientists suspect exists, because its possibility can account for otherwise unaccountable observations about the universe. Its central characteristic is that it is responsible for an acceleration in the universe's rate of expansion. According to this scenario, not only is the universe expanding, but its expansion is speeding up, and dark energy is the reason why. This would mean that eventually, bodies outside what is called our 'super cluster' (relatively local galaxies) would move so far away from us and at such a speed that they would become invisible. Our 'observable universe' then would empty out, and we would observe truly empty space.

While this Big Drift would leave our own Milky Way unscathed, an even darker interpretation of dark energy suggests that it would continue to operate, eventually dominating all other forces in the universe, even gravity. This would mean that galaxies and solar systems themselves would fall apart, their cohesion dissolved by the ubiquitous dark energy. But it won't stop there. Dark energy could even corrode the electrical and nuclear forces holding atoms together, disintegrating everything in what is known as the Big Rip, as if a loose thread hanging from reality was pulled until the whole fabric of space and time was undone.

Many worlds

Admittedly, all these possible ends of the worlds are a long way off, and most likely we will not be around in any way or form to observe them. Climate change, global warming, overpopulation, loss of natural resources, or even a rogue state may start the process of our species checking out long before the universe does. But if the notion of a Big Crunch, a Big Drift, or a Big Rip keeps you awake at night, science can offer some consolation. Throughout this book I've mentioned what is known as the anthropic cosmological principle, which, in its simplest form, suggests that our universe is one in which intelligent life like our own can appear – which seems rather obvious, since I am here writing about it and you are here reading about it. A stronger version suggests that our universe is one in which intelligent life like ourselves *had* to appear. There are just too many 'cosmic coincidences' necessary for living beings like ourselves to have appeared, for it to be the case that life 'just happened', Jacques Monod's Monte Carlo game notwithstanding. We seem to live in a 'Goldilocks' universe, with conditions being 'just right' for us to turn up.

Years ago, I worked as a science writer for the University of California, Los Angeles, and I interviewed one of the department's astronomers who worked on this question. One of the 'just rights', he told me, were gas giants like the planets Jupiter and Saturn, familiar figures in our solar system but apparently extremely rare elsewhere. Jupiter and Saturn, he explained, were important for life forming on the earth, because they shield our planet from meteor bombardments. Jupiter, he said, is 'a garbage man of the solar system, shielding us from objects that would have hit us'.[15] There are many other such 'just rights'. The expansion of the universe itself is one of them. If the gravitation had been just slightly stronger in the early universe, the Big Crunch would have happened already and neither we nor anything else would be here. But if it was just slightly weaker, then stars and galaxies might not have formed, and again, we wouldn't be here. Our distance from the sun is another 'just right'. A little bit closer and it would be too hot for life, a little further and it would be too cold. But just like Goldilocks' porridge, where we are is 'just right'. Many books have been written spelling out the various other 'just rights' – John Gribbin's *Cosmic Coincidences* (1989) is a good start – but the basic idea is that, as Danah Zohar remarks, 'Out of all the possible values the physical constants of the universe might have had, they seem, improbably, to have arrived at the narrow range of constants that favour the existence of carbon-based life forms'.[16] That is, us.

Understandably, not all scientists are happy with the idea that we live in a universe that seems somehow designed to produce us. In the first place it means we are not the chance occurrences that traditional science has said we are – not a lucky hit on the cosmic Monte Carlo game or some of Lovecraft's accidental 'rearrangements of atoms', but real life natives who really belong here: you can call us an indigenous species. And in the second place, if these cosmic coincidences are *not* simply coincidences, then the unpalatable idea that some intelligence is responsible for arranging them can raise its unwanted head. Did someone or thing *want* to produce us, and so made sure everything was just right for that to happen? (Are we, perhaps, a kind of synchronicity?) Swedenborg believed that the reason we exist is so that heaven can be populated. Angels, he believed, were originally human, and without us and beings like us – he believed other planets had life – heaven would be empty. The entire universe, according to Swedenborg, was created so that human beings, from earth and from other planets, could serve their cosmic purpose by becoming angels and moving on to heaven. It doesn't get more anthropic than this. Along with being a religious thinker, Swedenborg was a scientist, one of the most brilliant of his time, according to his countryman and Nobel Prize winner Svante Arrhenius.[17] So while we may take his ideas about an anthropic universe with some grains of salt, we can't just dismiss them *tout court*.

One way scientists unhappy about an anthropic universe get around the problem of cosmic coincidences, is through the concept of many worlds, associated with quantum theory. In short, the idea is that our universe is only one of countless others, existing in other dimensions. All possible other universes exist, but ours just happens to be the one in which we do, and in which the cosmic coincidences necessary for us to be here took place. Out of all the possible arrangements of forces, energies, and whatever else is needed to make a universe, in ours it just happens to be the case that those forces and energies combined in just such a way as to produce just such a universe that could result in just such beings as ourselves. And the fact that we are here is actually proof of this, as we could only have arisen in a universe that enjoyed 'the narrow range of constants that favour the existence of carbon-based life forms'. Our universe happens to be that one, so it shouldn't be surprising that we are here. Our universe is one that can produce us. In all those other universes that can't, we are not.

While the idea that our universe is only one of an infinite number may console some of us when we contemplate the cosmic fate science says is in store, I have to say that personally, the notion is not appealing. Aside from the circular reasoning involved in this desperate measure to avoid any notion of our being here 'on purpose', the 'many worlds' theory strikes me as suffering from the same flaw as Francis Crick's molecules or John Gray's drift. Just as Crick's molecules 'just happened' to have hit on the truth – that our molecules are the 'really real' things behind our thoughts, including those about molecules – so, in the 'many worlds' theory, our universe just happens to be the one in which beings like ourselves can come up with a theory of many worlds, in order to avoid thinking there is some reason for them being here.

One world

Clearly, just as with strings or superstrings, there is no way to test this theory, at least so far. And as far as I know, no one has seen any of these other universes in which we don't exist. But what troubles me about this is not that I can't test it or can't see it, but that once again, science has arrived at another way of avoiding any idea that this world, here and now, with us in it, is important. If this world, in which I struggle and am responsible and am obliged to make decisions, is only one world among many, perhaps an infinity of other worlds, then it seems to me, my actions and choices here are not that significant. If there are other worlds, the difference between this one and another one may not be as vast as the difference between a world that can sustain human life and one that cannot. The differences can be less drastic. Given the possibility of other worlds, there is no reason why there can't be other worlds in which human life does appear, but appears differently than it does here. As Danah Zohar writes: 'Quantum many-worlds theory argues that *every* possible sort of universe exists'.[18] So there may be a world in which I am a saint, and another in which I am a sinner. There may be a world in which I save lives, and another in which I destroy them. There may be a world in which I sacrifice my vital and pleasure values and deficiency needs in order to actualise the meta-need of pursuing a spiritual path, and a world in which I wallow in fleshpots and generally waste my time. Some may think such an idea is good, that it allows for greater variety, possibility, and multiplicity, and that the idea of only one world is severely limited. But I don't see it that way. What strikes me is that in such a many-worlds scenario, any good act I perform in this world may be cancelled out by a bad one performed in another, rather as matter is cancelled out by anti-matter, or the sparks of divine light are blotted out by the *klipoth*. If my efforts to repair the universe here are cancelled out by actions that damage it in another, what's the point? If I am a caretaker in this cosmos but a vandal in another, what difference do my actions make? If in all the possible universes in which I can exist, I live out all the possible scenarios of my life, why should I do anything in particular here? If I don't do the good that I know here, I will do it in another world, so there's nothing to worry about (although I may think the same thing in the other worlds, too, and so it doesn't get done there either). I might as well save myself the trouble and take it easy. Somewhere among the infinite worlds the good will get done, so relax.

Such a belief, it seems to me, subtracts any urgency from this life, in this world – subtracts, in fact, any value from it. One of the most dangerous ideas we can entertain is that our actions do not matter. This, combined with the notion that 'someone else will do it', leads to the very 'fallacy of insignificance' this book is trying to undermine. It leads to the 'unbearable lightness of being', that Milan Kundera wrote a novel about. What I want instead is the opposite. What I crave, and what I believe any self-actualising caretaker of the cosmos craves, is the *weight* of being, its resistance, something that drifts away if my actions and choices are cancelled out by their opposites. It is only in a world of limited possibilities that my choices can matter, because it is only in such a world that there is any risk and that anything can be at stake.

The invisible Earth

Someone who understood this was the poet Rilke. In the Ninth of his tremendous *Duino Elegies* (1923) – tremendous in the original sense of something that makes us tremble – Rilke writes:

> *Once* for each thing. Just once; no more. And we too,
> just once. And never again. But to have been
> this once, completely, even if only once:
> to have been at one with the earth, seems beyond undoing.[19]

Earlier in the poem Rilke writes that '*truly* being here is so much; because everything here apparently needs us, this fleeting world, which is some strange way keeps calling to us'. The world *needs* us because it is through us that it can be saved. In fact, Rilke's Ninth Elegy encapsulates the whole philosophy of 'cosmic caretaking' in so dense and powerful a manner that reading the poem and understanding it can be overwhelming. We save the fleeting things of the world, Rilke tells us, by drawing them *into ourselves*. 'Perhaps we are *here*,' Rilke tells us, 'in order to say: house, bridge, fountain, gate, pitcher, fruit-tree, window ... But to *say* them, you must understand ... *more* intensely than the Things themselves ever dreamed of existing.' The connection between Rilke's 'saying' and that of Heidegger and Saint-Martin is clear. What this saying accomplishes, Rilke tells us, is that it 'interiorises' the world, it brings it *inside* us, and makes it 'invisible'. 'Earth, isn't this what you want: to arise within us *invisible*?' he asks. This act of 'interiorising' is the essence of what Rilke called *Herzwerk*, 'heart work', and he explained what he meant by this in a letter to his Polish translator Witold von Hulewicz. (Readers of *The Quest for Hermes Trismegistus* will be aware of this and must, I am afraid, endure some repetition.)

Like many in the early twentieth century, Rilke believed that the rationalist-reductive view of life was emptying the world of meaning. This was especially true of the mass-produced junk rolling off countless assembly lines. The everyday things of life were increasingly being replaced by what Rilke called 'pseudo things' and 'Dummy life'. Where before everyday things, however inanimate, had some soul life, now they were inert. 'Even for our grandparents', Rilke writes, 'a "House", a "Well", a familiar tower, their very dress, ... was infinitely more intimate'.[20] In order to save these things from complete oblivion, Rilke said that we must take them into ourselves, and transform them into the furnishings of our interior world. By doing this, Rilke tells von Hulewicz, we become the 'bees of the invisible'. 'Our task is to stamp this provisional, perishing earth into ourselves so deeply, so painfully and passionately, that its being may arise again, "invisibly" in us.'[21] (We remember that earlier Rilke spoke of 'an interior that I never knew of.') We do this, Rilke tells us, not only for ourselves or only for the things, but for what Rilke calls 'the Whole'.

One possible result of this existential salvage operation, Rilke tells us, is that by transmuting the 'visible and tangible into the invisible vibration ... of our own nature' we may introduce 'new vibration-numbers into the vibration-spheres of the universe' – and we remember that string theory is about just such vibrations or oscillations. And 'since the various materials in the cosmos are only the results of different rates of vibration' – again, something science itself tells us – 'we are preparing in this way, not only intensities of a spiritual kind, but – who knows? – new substances, metals, nebulae, and stars'.[22] By taking the world into ourselves and making it invisible, Rilke tells us that we may actually be creating *new worlds*, somewhere in the depths of space. Rudolf Steiner said that the thoughts of people today will determine the physical form of the earth in the future. Rilke seems to agree – synchronistically, they were both working on these ideas at around the same time – but he takes it a bit further. By saving the world around us now – by performing *Herzwerk* and bringing the world 'inside' – we create new worlds. Not in the simplistic new age or post-modern sense in which reality is up for grabs because for them at bottom there is no objective reality. Reality is not amenable to our whims, but *something that makes demands on us*. By drawing the world into the black hole of our consciousness – it is a 'black hole' for the same reason a black hole itself is, because it is invisible – we may be creating somewhere out in space a 'white gusher', the proposed *other end* of black hole, where all the matter sucked into it comes bursting forth in new forms – nebulae and stars. If, as science tells us, there is a much larger 'unobservable' universe beyond our 'observable' one, could it also be true that there is a much larger 'unobservable mind' beyond that part of our inner world we can observe? I can look into myself only so far until I reach a kind of limit. I cannot immediately observe what is beyond that limit, but I know *something* is. 'I' seem to emerge like a fountain gushing out of a 'nowhere' that is nonetheless within me. It is as if I reach a kind of horizon, beyond which I cannot see. But I can see 'things' – ideas, thoughts, feelings, images – rising up from below that horizon, from, we could say, an inner underworld. I cannot have direct experience of whatever it is that is bringing those things up to the surface, but I trust it is there because I can see its effects. Perhaps in that inner 'event horizon' there is a place where the unobservable mind and the unobservable universe meet? I do not know, but if such were true, it could help us to understand what Rilke is saying.

The eighth day of creation

So, we may not need to be so concerned about the universe's 'heat death' or its inevitable demise through a Big Crunch, Big Drift, or Big Rip. What we do now with our consciousness may change all that. You and I, grappling with these thoughts, may in fact be saving the universe from extinction, or at least adding enough to it to compensate for whatever it loses to entropy or to dark energy. And although Bertrand Russell and many others believe that the second law of thermodynamics is unbreakable – for Sir Arthur Eddington, if you question it, you have no choice but to 'collapse in deepest humiliation' – not everyone agrees, and not all who have second thoughts are merely lonely poets.[23] In *What is Life?* (1944) the physicist Erwin Schrödinger – famous for his much abused cat – proposed the idea of 'negative entropy', or 'negentropy', to characterise life's apparent abrogation of this unbreakable second law.[24] More recent ideas about 'complexity' and 'self-organisation' seem to support this.[25] Life *is* organised energy, organised matter, and it seems to move in a direction *against* the general flow of matter; we can say that life flows *uphill*. This is what Bergson means by a 'creative evolution', one that *intentionally* invades matter and organises it for its own aims and purposes. These, Bergson believed, were to gain greater and greater control over matter and hence increased freedom. I suspect scientists would say that I misunderstand entropy, but I do believe that life is anti-entropic. The smallest amoeba moves against the downward stream, and since achieving this tenuous hold on matter, life has increased its control and its freedom, until today, it confronts a universe destined for oblivion with the most powerful anti-entropy weapon of all: the human mind. This is why Whitehead saw life as an 'offensive, directed against the repetitious mechanism of the universe'. For Whitehead and Bergson the fundamental character of life is that it is creative. The same is true for Berdyaev, Cassirer, and Scheler. When we 'complete' the world, when we 'represent' the 'unrepresented', when we infuse dead matter with meaning, when we fill the empty forms of reality with the living force of the imagination, we are moving against the tide that is carrying the fallen, physical world into nothingness. Each act of imagination, each moment of creative life stands up to the entire material universe and affirms the reality of meaning against the corroding solvents of entropy, dark matter, or whatever else may be dragging the physical world into oblivion. We are still, and will continue to be for some time, taking part in what Berdyaev calls 'the Eighth Day of Creation', the ongoing work of *tikkun*, taking care of and helping to create the cosmos, to transform it from a bleak, meaningless event into a living world. This means to humanise it, not in the small sense of reducing its vast, mysterious otherness to the triviality of the only human, but to release its sleeping life, its hidden meaning by acknowledging ourselves as the answer to its riddle.

Some proponents of an anthropic universe see our future in it as agents of a technology able to spread the human seed throughout the galaxies. Such seems the notion behind the Omega Point idea of Frank Tipler, one of the co-authors of *The Anthropic Cosmological Principle* (1986).[26] I cannot argue with Tipler's science, but I do know that for me the universe already is anthropic, in the sense that Swedenborg, Blake, Hermeticism, Kabbalah and other spiritual teachings present it. It does not need to be seeded by intergalactic computers spreading human intelligence throughout its depths.

The intelligence is already there. The beachheads and front lines of its advance are all around us. We do not need to travel to the far flung reaches of space to find them. We are at the barricades every morning. What is needed is a change in our consciousness, in our understanding of who and what we are.

First citizens of the cosmos

In the last decade of his life, Max Scheler developed an idea that is at one with the theme of this book. With Nikos Kazantzakis, Scheler came to believe that rather than an omnipotent, omniscient creator God, lording it over the universe, God, or the spirit, was actually weak. [27] As David Lindsay said, speaking of the muspel-fire, it was fighting for its life. For Scheler, God, the cosmos, and Man were all part of one effort of becoming. None were finished or fixed or perfect, but all were involved in an ongoing effort of realisation. With Gustav Fechner, Scheler believed that spirit needs matter, in order to become actual, in order for its values and meanings to become concrete. But spirit is weak and matter is strong and the effort of actualisation requires constant renewal, as Maslow's self-actualisers know only too well. On its own, spirit is free but ineffective. On its own, matter is enslaved but actual. It is only through the union of the two that any development, any evolution is possible. Life is that union, and as far as we know, it has made its most significant advance in us. Whether we live up to this reality or not is another question. An important one, clearly, but our failure to fulfil the task given to us is no argument against that task itself.

'In order to realise ... its inherent plenitude of ideas and values,' Scheler writes, 'the Ground of Being [the *Ein-Sof*] was compelled to release the world-creative drive. It was compelled ... to pay the price of this world process in order to realise its own essence in and through the temporal process'.[28] In other words, in order for God to become something more than just a good idea, He (She or It) had to plunge into the tough, thick, tangled jungle of matter and master it. Just as any creator, in order to create, has to give up the freedom of the ideal plane for the resistance of the actual one, God or Spirit had to relinquish freedom in the higher spheres for the difficulties of this one. Actually, to be more precise, it is only spirit or its embodiment in some living form, that experiences this world as difficult. Things aren't hard for a rock, if I can put it in that way. A cloud has no worries. It is only we and beings like ourselves that find this world difficult, because we are straining within it, trying to infuse it with more life. It is rather as if in coming to this planet we all had to wear clothes that were too tight, and we find ourselves constantly wanting to burst their seams. Spirit is that within us that needs to break free of what encumbers it, and it is only spirit that finds things uncomfortable. But it is not in our everyday 'good enough' selves that this realisation of Spirit takes place. Nothing is realised by taking things for granted and accepting the 'natural standpoint'. It is only by questioning this and asking who we really are, that we discover ourselves and our place in the universe.

It is through this discovery that we come to recognise that we are, in a sense, in the cosmos but not of it. We are, in fact, as Scheler argues, 'outside' the cosmos, not in a physical but in a metaphysical sense. Earlier I quoted Scheler's belief that animals always say 'Yes' to the cosmos, to life, to nature. This is because they are unable to step back from the press of things and 'objectify' it, turn it into an object of contemplation, something, as far as we can tell, only humans can do. We are, Scheler says, 'open to the world' and in this sense Man is a 'spiritual being ... that surpasses himself in the world'.[29] And it is only by making this separation, this surpassing, that we can discover who we are. As soon as we separate ourselves from the rest of nature and begin to look at it as an *object*, we begin to wonder about ourselves. "Where do I fit in in all this?" we wonder. "What is my place in the universe?" Yet that which us allows us to do this, that which makes it possible for us to turn the things of the world, and even the world itself, into an object, cannot itself be an object. Spirit can *never* be an object. Spirit as spirit is always an 'I', always, in a sense, a 'person', and as Berdyaev and others have noted, personality is the most mysterious thing in creation. We know this intuitively and it is why we find it wrong to treat other people as things. We are even beginning to feel this way toward animals, recognising in them an interiority that compels our sympathy – our 'with feeling'. And of course many 'primitive' or 'indigenous' people have for centuries felt the life in all life, in stones, trees, animals, something we touched on when discussing Cleve Backster's 'primary perception'.

Our ability to 'objectify' reality is what made it possible for us to separate from the great surround, to develop as independent beings, and to actualise our freedom. Now we also know that it is what has led, in many ways, to the abuse of the world around us, and also of other people. We cannot go back to a time when we did not have this power. And in any case, the cosmos itself doesn't want us to, because it needs us to save it. It is only by separating from the whole that we can ask who we are, and discover our true relationship to the whole. What that relationship is, has been the focus of this book. What we can do now is to try to understand our power, try to use it responsibly, and actively work at developing it so we can take the next step in our evolution. By doing this we become, as Scheler said, the 'first citizens of creation rather than its lord and master'.[30] And when that happens, we will all be caretakers of the cosmos.

Notes

Introduction: Saving the Universe

1. See Gary Lachman, *The Quest for Hermes Trismegistus* (Edinburgh: Floris Books, 2011) pp. 28–34.
2. See Elaine Pagels' classic work *The Gnostic Gospels* (New York: Random House, 1979) and Richard Smoley *Forbidden Faith: The Gnostic Legacy* (New York: Harper Collins, 2006) for an overview.
3. See my *Jung the Mystic* (New York: Tarcher/Penguin, 2010) and *In Search of P.D. Ouspensky: The Genius in the Shadow of Gurdjieff* (Wheaton, Ill: Quest Books, 2006).
4. Colin Wilson, *The Stature of Man* (Boston: Houghton Mifflin, 1959).
5. See *The Nag Hammadi Library in English* ed. James Robinson (New York: HarperOne, 2000).
6. Clement Salaman, Introduction to *Asclepius The Perfect Discourse of Hermes Trismegistus* (London: Duckworth, 2007) pp. 58, 61.
7. *The Dedalus Book of the 1960s: Turn Off Your Mind* (Sawtry, Cambs: Dedalus, 2010); US edition *Turn Off Your Mind: The Mystic Sixties and the Dark Side of the Age of Aquarius* (New York: Disinformation, 2003).
8. See my *A Secret History of Consciousness* (Great Barrington, MA: Lindisfarne, 2003) pp. 75–8, and *Rudolf Steiner: An Introduction to His Life and Work* (New York: Tarcher/Penguin, 2007) pp. 94–7.
9. This is the title of the American edition, noted above. The original English title, *The Age of Defeat*, was considered too depressing for American readers. A revised edition of *The Age of Defeat*, with a new introduction by Wilson, is available from Paupers Press http://pauper.stormloader.com/

1. The Other Side

1. The following account is taken from Sanford L. Drob's *Kabbalistic Metaphors* (Northvale, NJ: Jason Aronson Inc., 2000). I am indebted to Mr Drob's book, as well as his excellent website, for his interpretation of Luria's ideas.
2. Walter Benjamin 'Theses on the Philosophy of History' in *Illuminations* (London: HarperCollins, 1992).
2. See my *The Dedalus Book of Literary Suicides: Dead Letters* (Sawtry, Cambs: Dedalus, 2008) pp. 100–09.

4 See Peter Demetz's Introduction to Walter Benjamin *Reflections* (New York: Harcourt, Brace, Jovanovich, 1978) pp. xxiv–xxvi
5 Drob, p. 11.
6 Ibid.
7 This means that it would take more than six times as long as the universe has been in existence for a beam of light to travel from one side of the observable universe to the other.
8 But this, as Wittgenstein tells us, is the mystical. *Tractatus Logico-Philosophicus* (London: Routledge Kegan Paul, 1961) p. 149, 6.45, 6.5.
9 Andromeda itself is on a collision course with us, a head-on that will occur four billion years from now. http://www.nasa.gov/mission_pages/hubble/science/milky-way-collide.html
10 http://www.nytimes.com/1997/01/21/science/physicists-confirm-power-of-nothing-measuring-force-of-universal-flux.html?pagewanted=all&src=pm
11 http://www.guardian.co.uk/science/shortcuts/2012/jul/02/higgs-boson-found-definite-maybe
12 George Steiner, ed. *Is Science Nearing its Limits?* (Manchester: Carcanet, 2008) p. xxix.
13 The strange thing is that anyone thought this was news. See my article 'This Just In: God is Dead (Again) in *Fortean Times* 272 March 2011 p. 58.
14 Steven Weinberg *The First Three Minutes* (New York: Basic Books, 1993) p. 154.
15 http://physicsworld.com/blog/2011/01/the_theory_of_everything_on_a.html
16 See my *A Secret History of Consciousness* (Great Barrington, MA: Lindisfarne, 2003) pp. xxi–xxvii.
17 Jacques Monod, *Chance and Necessity* (New York: Vintage Books, 1971) p. 146. Monod, of course, is not alone in this belief. According to the American palaeontologist George Gaylord Simpson, whose view can be taken as standard, 'Man is the result of a purposeless and materialistic process that did not have him in mind. He was not planned.' Quoted in Arthur Koestler *The Ghost in the Machine* (New York: Macmillan, 1967) p. 151.
18 Peter Caws *Sartre* (London: Routledge, 1999) p. 5.
19 Jean Paul Sartre *Being and Nothingness* (Abingdon: Routledge, 2000) p. 615.
20 H.P. Lovecraft *Selected Letters 1925–1929* (Sauk City, WI: Arkham House, 1968) p. 41.
21 S.T. Joshi *H.P. Lovecraft: A Life* (West Warwick, RI: Necronomicon Press, 1996) p. 642.
22 Readers wanting an idea of what Sartre's 'nausea' is like can get a sense of it by repeating a word over and over; after a time it seems to lose its meaning, and becomes simply a sound emerging from our mouths.
23 Jean Paul Sartre *Nausea* (Harmondsworth: Penguin Books, 1975) p. 182.
24 Simone de Beauvoir *The Prime of Life* (New York: Perseus Books, 1992) pp. 168–170.
25 H.P. Lovecraft, in *Tales of the Cthulhu Mythos*, ed. August Derleth (Sauk City, WI: Arkham House, 1969) p. 39.
26 Ibid. August Derleth, 'The Cthulhu Mythos', p. vii.
27 Strangely, another popular literary work of 'cosmicism' aims not to frighten its readers, but to make them laugh. Its humour, however, is bleak. Douglas Adams' *The Hitchhiker's Guide to the Galaxy* (1978) uses the same trope of a uni-

verse too big for human life to have any meaning within it. The story begins with the earth being demolished in order to make way for a galactic by-pass. The rest of the jokes are of the same deflating character.
28 H.G. Wells *The War of the Worlds* (London: Penguin, 2005) p. 7.
29 Colin Wilson *The Strength to Dream* (Houghton Mifflin Co.: Cambridge, 1962) pp. 1–10.
30 S.T. Joshi *The Weird Tale* (Austin, TX: University of Texas Press, 1990) pp. 202–03.
31 John Gray *Straw Dogs: Thoughts on Humans and Other Animals* (London: Granta, 2002).
32 Gary Lachman 'We're Only Human' *Fortean Times* 193 February 2005 p. 54.
33 See my *Turn Off Your Mind: The Dedalus Book of the 1960s* (Sawtry, Cambs.: Dedalus, 2010) p. 326.
34 Charles Manson *The Manson File*, ed. Nicholas Schreck, (New York: Amok Press, 1988) p. 109.
35 In 1975, Fromme was imprisoned for the attempted assassination of President Gerald Ford. She claimed she made the attempted in order to 'save the redwoods'.
36 Nicholas Goodrick-Clarke *Hitler's Priestess: Savitri Devi, the Hindu Aryan Myth and Neo-Nazism* (New York: NYU Press, 1998).
37 Dave Foreman 'Whither Earth First!' *Earth First!* No. 1, November 1987.
38 Gray 2002, p.120.
39 Freud, too, believed we were sick, or at least incurably neurotic, the cause of our illness being civilisation, which forces us to inhibit our natural instincts, a theme which goes back to Rousseau and the myth of the 'noble savage'. Much of the Romanticism of the nineteenth and twentieth centuries echoes this notion; writers such as D.H. Lawrence and Henry Miller, just to name two, express it admirably. In *The Ghost in the Machine* (1967) Arthur Koestler argued for a remedy to the 'schizophysiology' of our nervous systems, the imbalance between our intellectual and emotional selves, that gives rise to the 'paranoid streak' in humankind. I mention these to show that Gray's diagnosis is not uncommon.
40 http://newhumanist.org.uk/661/escape-from-eden
41 Kathleen Raine *The Inner Journey of the Poet* (New York: George Braziller, 1982) pp. 7–8.
42 John Gray, *The Immortalization Commission* (London: Penguin, 2011) p. 20.
43 For a time in the 1930s, Cioran was involved with the notorious Legion of the Archangel Michael in his native Romania, as was his close friend Mircea Eliade. See my *Politics and the Occult* (Wheaton, Ill: Quest Books, 2008) pp. 226–27. Cioran later lamented his involvement, unlike Eliade.
44 Julian Huxley *New Bottles for New Wine* (London: Chatto & Windus, 1957) pp. 13–17.
45 Gray 2002, p. 6.
46 Gray 2011, p. 40.
47 Ibid. pp. 17–18.
48 Ibid. p. 47.
49 Ibid. p. 126.
50 Francis Crick *The Astonishing Hypothesis: The Scientific Search for the Soul* (London: Simon and Schuster, 1994) p. 3.
51 Raine 1982, pp. 3–4.
52 Colin Wilson *Alien Dawn* (London: Virgin Publishing, 1998) p. 245.

2. A Dweller on Two Worlds

1. See *The Road to Reality* (New York: Knopf, 2005).
2. The fragmentary character of our times has recently received a kind of official declaration, although the people responsible for this may not be aware of it. On 5 July 2012 an imposing new skyscraper was 'unveiled' in London. The Shard, as it is called, stands some 1016 feet above the city. According to the Oxford Dictionary, a 'shard' is a 'broken piece of pottery or glass'. The 'tallest building in Western Europe' turns out to be a fragment. http://www.dailymail.co.uk/travel/article-2169601/The-Shard-laser-lights-London-declared-tallest-building-Europe.html?ito=feeds-newsxml
3. Iain McGilchrist *The Master and His Emissary* (New Haven: Yale University Press, 2010) p. 442.
4. Kathleen Raine *Defending Ancient Springs* (West Stockbridge, MA: Lindisfarne, 1985) p. 156; Jacques Barzun *The Use and Abuse of Art* (Princeton: Princeton University Press, 1974) p. 57. Kathleen Raine's essay 'The Use of the Beautiful' was first published in 1967.
5. Barzun 1974, p. 58.
6. McGilchrist, p. 443.
7. Nikolai Berdyaev, *The Meaning of the Creative Act* (New York: Collier Books, 1962) pp. 209, 228.
8. Ibid. p. 132.
9. Lachman 2011 p. 66.
10. See Lachman 2003, pp. 104–105.
11. The birth of the modern, scientific world, was predicated on the death, or at least banishment, of the ancient, animistic one. See Lachman 2011, pp. 181–84.
12. Lachman 2011, pp. 183–84.
13. Friedrich Nietzsche, *The Will to Power* ed. Walter Kaufmann (New York: Random House, 1967) p. 8.
14. S. Foster Damon *A Blake Dictionary* (Boulder, CO: Shambhala, 1979) p. 335.
15. Emanuel Swedenborg *The True Christian Religion* (London: Swedenborg Society, 1988) §767.
16. For more on Milosz see my *Politics and the Occult* (Wheaton, IL: Quest Books, 2008) pp. 187–89, and *The Dedalus Book of the Occult: A Dark Muse* (Sawtry, Cambs: Dedalus, 2003) pp. 243–251.
17. O.V. de Lubicz Milosz *The Noble Traveller: The Life and Writings of O.V. de L. Milosz* ed. Christopher Bamford (West Stockbridge, MA: Lindisfarne, 1985) p. 246–47.
18. Philip Sherrard *Human Image: World Image* (Ipswich: Golgonooza Press, 1992) p. 134.
19. Adam Strange first appeared in *Showcase* no. 17 November–December 1957, published by DC Comics; he moved to *Mystery in Space* in 1959. I started reading comics at around 5 years old, so my first encounter with the series must have been in 1961; I was born in 1955. Julius Schwartz, the editor, said that he named the hero after the original Adam.
20. R.J. Hollingdale's translation of the *Tales of Hoffmann* (London: Penguin, 1982) is an excellent collection.
21. Steven Pinker *How the Mind Works* (New York: Norton, 1997) pp. 525–32.

I would agree with Pinker that muzak and much popular music is a kind of 'auditory cheesecake', but Beethoven and Co. are clearly much more nourishing fare.
22 *Thayer's Life of Beethoven Vol. I* revised and edited by Elliot Forbes (Princeton: Princeton University Press, 1991) p. 496.
23 Readers can find translations of Hoffmann's tale, as well as Goethe's 'Fairy Tale of the Green Snake and Beautiful Lily' in *The Dedalus Occult Reader: The Garden of Hermetic Dreams* (Sawtry, Cambs: Dedalus, 2004) which I edited. They can also find more about Hoffmann in my essay on him in *The Dedalus Book of the Occult: A Dark Muse* (Sawtry, Cambs: Dedalus, 2003) pp. 76–82, where I discuss in more detail Hoffmann's 'serapiontic principle', which argues for the necessity of magic to be *contrasted* with the everyday world. They can also find essays on Goethe, Novalis and other Romantic writers, and their relationship to the western esoteric tradition.
24 Colin Wilson *The Haunted Man: The Strange Genius of David Lindsay* (San Bernadino, CA: Borgo Press, 1979). p. 9.
25 David Lindsay *The Haunted Woman* (Edinburgh: Canongate, 1987) p. 23.
26 Raine 1985, p. 113.
27 Lachman 2011, pp. 34–41.
28 In *An Essay on the Origin of Thought* (Columbus: Ohio University Press, 1974) the late Danish philosopher and musician Jurij Moskvitin speaks of an 'Anthroposphere' , which is in many ways similar to Husserl's 'communal life world' but is also along the lines of Karl Popper's 'World 3'. See Lachman *A Secret History of Consciousness* (Great Barrington, MA: Lindisfarne, 2003) pp. 179–80. See also p. 298, section 21, note 1.
29 Lachman 2003 p. 71.
30 Hermann Hesse *The Journey to the East* (London: Picador, 1995) p. 13.
31 Although in 1875 Madame Blavatsky founded the Theosophical Society in New York, the term 'theosophy' had a long history prior to this. It was used, for example, in the metaphysical and alchemical writings of the Silesian cobbler Jacob Boehme (1575–1624), which were a central influence on Saint-Martin.
32 For an outline of Saint-Martin's life see Lachman 2003 pp. 28–32.
33 Quoted in Ernst Benz, *The Mystical Sources of German Romantic Philosophy* (Allison Park, PA: Pickwick Publications, 1983) p. 75.
34 http://www.inner.org/worlds/reshimu.htm
35 George Steiner *Martin Heidegger* (Chicago: University of Chicago Press, 1987) p. 128.
36 George Steiner *Grammars of Creation* (London: Faber and Faber, 2001) p. 16.
37 McGilchrist 2010 p. 3.
38 See 'Under Ben Bulben'.
39 See, for example, Jung's ideas about what he calls 'the transcendent function' in my *Jung the Mystic* (New York: Tarcher/Penguin, 2010) pp. 115–19.
40 McGilchrist argues that this division of labour can be seen in non-human species as well, and instances birds, who, in order to eat, need to be able to pinpoint bits of food amidst inedible items, but also need to be aware of what is going on around them, so as to not become food themselves. This double duty is linked, he argues, to a asymmetry in avian brains which comes to a fuller flowering in our human cerebral hemispheres.
41 Interested readers may want to take a look at my review of McGilchrist's book

at http://lareviewofbooks.org/article.php?id=489
42 Colin Wilson *Poetry and Mysticism* (San Francisco: City Lights, 1970) pp. 53–82.
43 R.A. Schwaller de Lubicz *Esotericism and Symbol* (New York: Inner Traditions, 1985) p. 49.
44 R.A. Schwaller de Lubicz *Nature Word* (West Stockbridge: Lindisfarne Press, 1982) p. 135.
45 Henri Bergson *Introduction to Metaphysics* (New York: G.P. Putnam's Sons, 1912) p. 7.
46 I am indebted to Colin Wilson's discussion of Whitehead's *Symbolism: Its Meaning and Effect* in *Beyond the Outsider* (Boston: Houghton Mifflin Co., 1965) pp. 70–77 and *Below the Iceberg* (San Bernardino: The Borgo Press, 1998) pp. 118–125.
47 Alfred North Whitehead *Essays in Science and Philosophy* (London: Rider, 1948) pp. 72–74.
48 McGilchrist 2010 p. 22.

3. Doing the Good that You Know

1 Drob p. 223.
2 http://www.newkabbalah.com/stein.html
3 Apparently this 'planned failure' was itself not easy to effect. According to George Steiner, rabbinic tradition records that there were twenty-six aborted attempts before God actually managed to create the world. (Steiner, 2001 p. 26).
4 Drob, pp. 22, 25.
5 Ibid. p.150.
6 Max Scheler, *Man's Place in Nature* (New York: Noonday Press, 1978) p. 71.
7 P.D. Ouspensky *The Fourth Way* (New York: Alfred A. Knopf, 1970) p. 23.
8 Sefton Delmer *Weimar Germany: Democracy on Trial* (London: Macdonald/American Heritage, 1972) p. 122.
9 Nikos Kazantzakis, *The Saviours of God: Spiritual Exercises* (New York: Simon & Schuster, 1960) p. 4.
10 Kazantzakis' play was eventually completed and published in an English translation, *Buddha* (San Diego: Avant Books, 1983).
11 Kazantzakis 1960 p. 9.
12 Ibid. p. 5.
13 Ibid. p. 21.
14 Ibid. p. 54 An even more Hermetic expression is Kazantzakis' admonition to 'Train your heart to govern as spacious an arena as it can ... Train your eye to gaze on people moving in great stretches of time,' (p. 78) which is remarkably similar to Hermes Trismegistus' teaching to 'Command your soul to go anywhere, and it will be there quicker than your command' and to 'Be free from every body, transcend all time' as well as other similar dictums. See Lachman 2011 pp. 35–36
15 Ibid. pp. 92, 103, 105, 106.
16 David Lindsay *A Voyage to Arcturus* (Edinburgh: Canongate, 1992) p. 301.
17 Kazantzakis pp. 115, 68.

18 Victor Frankl, *Man's Search for Meaning* (New York: Simon & Schuster, 1984) p. 105.
19 Ibid. p. 134.
20 http://www.responseabilityalliance.com/html/statue_of_responsibility_found.html
21 http://www.newkabbalah.com/stein.html
22 It is a curious fact that many 'life affirming' philosophies grow out of difficult conditions, while an easy life seems to breed pessimism. Nietzsche suffered from a collection of physical and neurological complaints that kept him constantly in search of congenial climate. He was also the loneliest man in Europe. Yet he developed a philosophy of sheer affirmation. Schopenhauer, his erstwhile mentor, was as pessimistic a philosopher as can be imagined, embracing a Buddhistic denial of life. Yet he had a comfortable, cosy life, was relatively well off, and enjoyed sensual pleasures without guilt. It is as if the effort needed to endure difficult conditions pushes us into an appreciation of things from which we do not easily sink into taking them for granted.
23 Frankl p. 108.
24 Ibid. pp. 113–14.
25 McGilchrist p. 28.
26 Ibid.
27 Ibid. p. 7.
28 Daniel C. Matt *The Essential Kabbalah* (New York: Harper Collins, 1996) p. 16.
29 Gershom Scholem *On The Kabbalah and its Symbolism* (London: Routledge & Kegan Paul, 1965) p. 117.
30 Hesse 1995, p. 21.
31 Samuel Butler, *The Notebooks of Samuel Butler* (New York: E.P. Dutton & Co., 1917) p. 27.
32 Wilson Van Dusen *The Presence of Other Worlds* (London: Wildwood House, 1975) p. 213.
33 For cultural relativism, there are no 'extra-cultural' criteria by which we can judge whether or not a practice or belief is 'good'; all such criteria (except utilitarian ones) are cultural. Insofar as it enables people who are oppressed by one set of values to liberate themselves from it and to assert their own, we can see such 'relativism' as beneficial, as, indeed, 'good'. But what if these liberated values are, in the traditional sense, 'evil', as, say, the values of Nazi Germany were? Under Hitler, it was good to arrest, torture, and murder Jews as well as various other 'sub-humans'. Faced with this, is it still good to be a cultural relativist? Or is it good to oppose these goods, as most people would? Indeed, how can a cultural relativist argue about opposing goods? If all criteria are cultural, what does he use to judge the value of different goods? It is good to recognise that ideas about the good are cultural, but is that idea itself cultural? If it is, then there can be a culture which rejects it, and a cultural relativist can have no argument against this. It is good for Nazis to kill Jews, but non-Nazis say it isn't. What right do non-Nazis have to prevent Nazis from doing the good that they know? If we say that the Nazis' killing of Jews isn't really a good, but is motivated by hatred, fear, greed, and other 'negative' values, what do we oppose to these and why are they not 'good'? Of course I am not in any way suggesting that Nazi values were 'good', merely showing how difficult it would be for a cultural relativist to say they weren't. While it is clear that ideas about an 'absolute' good and evil are problematic, it is also clear that a relativis-

tic approach to these matters is itself rather shaky.
34 Abraham Maslow *The Farther Reaches of Human Nature* (New York: Penguin, 1976) p. 9.
35 Ibid. p. 6.
36 Van Dusen p. 213.
37 Richard Sennett, *The Craftsman* (London: Allen Lane, 2008) p. 9.
38 Maslow, 1976 p. 113.
39 Van Dusen p. 215.
40 Ibid.
41 See note 21 Chapter One.
42 Drob p. 65.
43 See my *Swedenborg: An Introduction to His Life and Ideas* (New York: Penguin, 2012) pp. 30–1 and throughout, for more on Swedenborg's erotic spirituality.
44 For more on Count Zinzendorf, the Moravians, and Rabbi Falk, see my *Politics and the Occult* (Wheaton, Il: Quest, 2008) pp. 53–65.
45 Ibid.
46 Berdyaev 1962 p. 168.
47 Vladimir Solovyov, *The Meaning of Love* (West Stockbridge, MA: Inner Traditions/Lindisfarne, 1985) p. 55.
48 Ibid. p. 60.
49 Ibid. p. 10.
50 Ibid. p. 105.
51 Ibid. p. 43.
52 Berdyaev 1962 p. 171.
53 Ibid.
54 In his novel *Strange Life of Ivan Osokin* (London: Arkana, 1987) P.D. Ouspensky expresses this when his young hero, after his first embrace with the girl Tanechka, says 'Tanechka is a part of nature, like this field, or the wood, or the river. I never imagined that the feeling of woman was so much like the feeling of nature' (p. 84) Ouspensky was another Russian fascinated with the cosmic power of sexual love. In *Tertium Organum* (New York: Alfred A. Knopf, 1981) he wrote, echoing Berdyaev, that 'love is a cosmic phenomenon' which is 'something quite different, and of a different order from the small events of earthly life'. It was, he said, 'the way to sanctity' (pp. 138, 289).
55 I am in fact indebted to McGilchrist's references to Scheler in *The Master and His Emissary* for drawing my attention to his work.
56 Max Scheler *The Nature of Sympathy* (London: Routledge & Kegan Paul, 1979) p. 127.
57 Ibid. p. 126.
58 Ibid. p. 11.
59 Ibid. p. 117.
60 Berdyaev 1962 p. 205.
61 Ibid. p. 115.
62 John Raphael Staude *Max Scheler* (New York: The Free Press, 1967) p. 6.
63 Herbert Spiegelberg *The Phenomenological Movement Vol. 1* (The Hague: Martinus Nijhoff, 1976) p. 262.

4. The Good Society

1. Max Scheler *Problems of a Sociology of Knowledge* (London: Routledge and Kegan Paul, 1980) p. 26.
2. See Peter Viereck's *Metapolitics: The Roots of the Nazi Mind* (Piscataway, NJ: Transaction Publishers, 2003).
3. Scheler, 1980 p. 26.
4. For example, see *A Secret History of Consciousness* in which I examine the similarities between the 'consciousness structures' of Jean Gebser and the 'spiritual ages' of Rudolf Steiner.
5. I should point out that by a 'direct perception' of another's mind neither I nor Scheler mean that we have a direct perception of the *contents* of that mind. Our perception is of that mind itself, that is to say, we are aware that others have minds, inner worlds, just as we do. But we are not directly aware of what that mind is thinking, feeling, or perceiving. Which is not to say that we cannot be. I believe telepathy is possible, and I think Scheler would think so too. But he did argue for an area of 'absolute privacy', an *Intimsphäre* 'impenetrable to other people's scrutiny'. (Spiegelberg p. 262).
6. In *Tintern Abbey* Wordsworth writes of a 'joy of elevated thoughts; a sense sublime/Of something far more deeply interfused/Whose dwelling is the light of setting suns'.
7. Max Scheler, *The Nature of Sympathy* (London: Routledge and Kegan Paul, 1979) p. xv.
8. Ibid.
9. George Steiner, *Has Truth a Future?* (London: BBC, 1978) pp. 16–17.
10. Maslow also agreed with Scheler on the cognitive power of love. In his clinical research, Maslow discovered that 'loving perception ... produced kinds of knowledge that were not available to non-lovers.' It produces 'interest and even fascination, and therefore great patience and long hours of observation', which reminds us of McGilchrist's 'patient and detailed attention to the world'. 'If we love or are fascinated or are profoundly interested, we are less tempted to interfere, to control, to change, to improve ... ' We can see whatever we are observing 'more truly as it is in its own nature rather than as we would like it to be ... ' Maslow called this approach 'Taoist objectivity'. (Abraham Maslow *The Farther Reaches of Human Nature*, New York: Penguin Books, 1976, pp. 16–17.) All of this seems pretty much in line with Scheler's ideas about phenomenological love (Chapter Three).
11. Steiner 1978, p. 3.
12. H.G. Wells, *Experiment in Autobiography* (London: Gollancz, 1934) p.16.
13. Even this contrast is not quite right, as 'pulled' still has an element of passivity, as an iron filing is 'pulled' by a magnet. The idea is that rather than being driven by forces acting upon us and reacting to them out of sheer necessity, this new arrangement is one in which we freely choose to pursue a goal that has nothing to do with the demands of necessity.
14. Maslow 1976, p. 5.
15. See C.G. Jung 'Adaptation, Individuation, Collectivity' in *Collected Works Volume 17* (London: Routledge, 1991). Also Lachman 2010 pp. 132–35.
16. Maslow 1976, p. 8.

17 Ibid.
18 Ibid. p. 11.
19 See Abraham Maslow *Maslow on Management* (Chichester, NY: John Wiley, 1998). Also Maslow 1976 pp. 227–28.
20 Ibid. pp. 19, 205.
21 A Jew, Maslow grew up facing anti-Semitism, and he had an unhappy relationship with his mother. See Edward Hoffman *The Right To Be Human: A Biography of Abraham Maslow* (New York: Tarcher, 1988).
22 Wells, of course, was not alone in this. In *Tertium Organum* (1912) P.D. Ouspensky suggested that a new type of person was emerging, one focussed on precisely the kinds of values that Maslow and Scheler speak of. This notion was also echoed by Hermann Hesse in his novel *Demian* (1919). In *Cosmic Consciousness* (1901) the psychologist R.M. Bucke argued that humanity was evolving into a higher, wider form of consciousness, an idea informing A. R. Orage's *Consciousness: Animal, Human, Superman* (1904). Nietzsche of course spoke of the *Übermensch*. It is not an idea widely accepted today.
23 Wells had explored this theme in earlier works such as *The Food of the Gods* (1904) and *In the Days of the Comet* (1906).
24 H.G. Wells, *A Modern Utopia* (London: Penguin Books, 2005) p. 112.
25 Ibid. p. 113.
26 Ibid.
27 Ibid. p. 92.
28 The full text is available online at: http://gaslight.mtroyal.ca/repsouth.htm and the reader can find more about Bryusov in my book *The Dedalus Book of the Occult: A Dark Muse* (Sawtry, Cambs.: Dedalus, 2003).
29 William James 'What Makes a Life Significant' full text at http://www.des.emory.edu/mfp/jsignificant.html
30 Ibid.
31 Abraham Maslow 'Humanistic Biology: Elitist Implications of the Concept of Full Humanness' in *Future Visions: The Unpublished Papers of Abraham Maslow* ed. Edward Hoffman (Thousand Oaks, CA: Sage Publications, 1996).
32 George Steiner voiced a similar concern. He worried about how the 'bad' idea that some intellectual talents and abilities may be limited to a few *naturally*, and not because of social or environmental factors, would be understood by society. Steiner 1978, p. 14.
33 Readers can find some thoughts about Maslow's idea at my web site http://garylachman.co.uk/2011/02/05/an-out-take-from-politics-and-the-the-occult/
34 Max Scheler *The Nature of Sympathy* trans. Peter Heath (London: Routledge and Kegan Paul, 1979) p. 35.
35 Berdyaev p. 262.
36 Ibid. p. 263.

5. Beyond Nature

1 McGilchrist p. 28.
2 Ibid. p.5.
3 Lachman 2003, pp. 217–231 and Lachman 2011 pp. 122–127.
4 Jacob Burckhardt *The Civilization of the Renaissance in Italy* (London: Phaidon

Press, 1944) p. 179).
5 Ernst Cassirer *The Individual and the Cosmos in Renaissance Philosophy* (Philadelphia: University of Pennsylvania Press, 1972) pp. 141–44).
6 James Hillman, *Re-Visioning Psychology* (New York: Harper Colophon Books, 1977) pp. 196–97).
7 From 'The Tables Turned', complete poem at http://history.hanover.edu/courses/excerpts/111word.html
8 Mary Wollstonecraft *Letters Written During a Short Residence in Sweden, Norway, and Denmark* (Harmondsworth: Penguin, 1987) p. 68.
9 Rainer Maria Rilke *Duino Elegies* in *The Selected Poetry of Rainer Maria Rilke* ed. and translated by Stephen Mitchell (New York: Vintage Books, 1984) p. 151.
10 Ibid. p. 65. For more on Mary Wollstonecraft and the Romantics see my article 'Mary Wollstonecraft and Romantic Consciousness' at: http://www.enlightennext.org/magazine/j37/lachman.asp. See also Lachman *The Dedalus Book of Literary Suicides: Dead Letters* (Sawtry, Cambs: Dedalus, 2008) pp. 223–230.
11 In *The Craft of the Novel* (1975) Colin Wilson suggests that the publication of Samuel Richardson's novel *Pamela* (1740) was the starting point of this change in consciousness. There were novels before this, *Don Quixote* (1605–1615) and *Robinson Crusoe* (1719) for example, but *Pamela* was something different, and is considered the first novel in the modern sense. *Robinson Crusoe* and *Don Quixote* were about 'far away places with strange sounding names', and told of extraordinary adventures in exotic lands. *Pamela* was about everyday life, and was an immense relief from the interminable boredom that most people must have felt, when they were not wrestling with nature in order to scratch out a living. For many people then, the only source of entertainment was the weekly sermon. To modern tastes Richardson's *Pamela* is an unbearably slow account of the eponymous heroine's defence of her virtue from the repeated and perpetually frustrated attempts on it by her employer. What made it different from *Robinson Crusoe* and earlier epics is that the readers could put themselves in Pamela's – or her would-be seducer's – shoes, and *imagine what their own lives would be like* if they were in her situation. (This, of course, is the appeal of every soap opera.) They could spend hours in an entirely new dimension, an inner landscape that the men and women who read the book were only then discovering. This is something we, who take reading for granted, find difficult to grasp, simply because we have grown up with it and with our own sense of having an 'inner world', we which take as a given. We have to go back to our own earliest experiences of the 'magic' of reading – or of listening to music – to get a sense of the tremendous change that had taken place in human consciousness.

Aptly enough, Richardson's writing of *Pamela* is an example of one of Maslow's 'means-activity' becoming an 'ends-activity', that I mention in Chapter Four. The book started out as a 'self help' text on how girls going into service could stay out of trouble (i.e. retain their virtue), and is written in the form of letters. But Richardson became so engrossed in the situation and the responses of his heroine that the story took on a life of its own. He himself wanted to know how things would turn out and how Pamela could ward off her suitor. From being a utilitarian activity with a particular aim in view, writing the work became for Richardson an end in itself. Likewise, although Richardson defended the work from criticisms of licentiousness by saying that it was educational – that it 'instructed through entertainment', an early version

of our own 'edutainment' – it was clear that its appeal to its readers was less instruction (utility) than sheer enjoyment.
12 Rainer Maria Rilke, *The Notebooks of Malte Laurids Briggs* trans. Stephen Mitchell (New York: Vintage Books, 1985) p. 5.
13 Gebser, pp. 15, 12.
14 Owen Barfield *Saving the Appearances* (New York: Harcourt Brace & World, 1957) pp. 76–77.
15 Quoted in Kathleen Raine *The Inner Journey of the Poet* (New York: George Braziller, 1982) p. 203.
16 Lachman 2003. See also my articles 'Jean Gebser: Cartographer of Consciousness at http://magazine.enlightennext.org/2011/01/26/jean-gebser-cartographer-of-consciousness/ and 'Owen Barfield and the Evolution of Consciousness' at: http://davidlavery.net/barfield/barfield_scholarship/lachman.html
17 Gebser receives some support for his dating in the work of the psychologist Julian Jaynes. In *The Origin of Consciousness in the Breakdown of the Bicameral Mind* (1976), Jaynes argues that until 1250 BC, human beings were not self-conscious, in the way we are. Their consciousness was 'bicameral', two-chambered, meaning that the communication between the right and left brain, which we take as natural, had not yet been achieved. Jaynes argues that until then, human beings heard voices in their head, telling them what to do. They interpreted these as the gods, but in fact they were listening to the right brain. See Lachman 2003, pp. 143–49.
18 Steiner's writings and lectures on the evolutionary stages of consciousness are varied and diverse, but the idea permeates practically all his work. *An Outline of Occult Science* (1909) is a comprehensive but often unwieldy account of human and cosmic evolution. *Cosmic Memory* (1939), a collection of essays originally published in Steiner's journal *Lucifer Gnosis* (1904) are more accessible; see pp. 196–208. I summarise Steiner's evolutionary ideas in Lachman 2007 pp. 125–151.
19 See Lachman 2003, pp. 232–264.
20 See Lachman 2011, pp. 180–19.
21 See my *Swedenborg: An Introduction to His Life and Ideas* (New York: Tarcher/Penguin, 2012) p. 91.
22 William Blake *The Complete Poetry and Prose of William Blake* ed. David V. Erdman (Berkeley: University of California Press, 1982) pp. 565–66.
23 Ibid. pp. 565, 555.
24 Raine 1982,, p. 197.
25 S. Foster Damon *A Blake Dictionary* (Boulder: Shambhala, 1979) p. 445.
26 Ibid. p. 451.
27 Raine 1982, pp. 192–93.
28 Blake 1982, p. 385.
29 Blake and Saint-Martin were contemporaries – Saint-Martin was about fourteen years older – and there is a slim possibility that they could have met. In 1787 Saint-Martin visited the London Theosophical Society. This was no relation to Madame Blavatsky's Theosophical Society, which was founded in New York in 1875, but a study group, which began in 1783, devoted to Swedenborg's work. Blake was a member of the society, and it is just possible he was present when Saint-Martin made his visit. If so and if they did meet, of the two, Blake was truly the unknown philosopher. Saint-Martin's work was

well known at the time, but, as mentioned, Blake was obscure.
30 A.E. Waite *The Life of Louis Claude de Saint-Martin* (London: Philip Wellby, 1901) p. 113.
31 Quoted in Ibid. p. 117.
32 Ibid. p. 118.
33 Ibid. p. 139.
34 Ibid.
35 Ibid. p .151.
36 Ibid. p. 190.
37 Ibid.
38 Ibid. 232.
39 Some animals can of course camouflage themselves, or appear to be something they are not. But I would not consider this lying in the sense that they consciously know that they are providing false information.
40 George Steiner *Martin Heidegger* (Chicago: University of Chicago Press, 1987) p. 95.
41 Ibid. pp. 95–96.
42 Waite 1901, p. 233.
43 Ibid.
44 Ibid. p. 186.
45 Bob Dylan 'If Dog's Run Free' on *New Morning* (Columbia Records 1970).
46 Heinrich von Kleist 'On the Marionette Theatre' in *Essays on Dolls*, ed. Idris Parry and Paul Keegan (London: Penguin, n.d.) p. 11.
47 Nikolai Berdyaev *The Destiny of Man* (London: Geoffrey Bly, 1937) p. 131.
48 Ibid. p. 132.
49 Ibid. pp. 46, 45.
50 Ibid. p. 11.
51 Ibid. p. 48.
52 Peter Tompkins and Christopher Bird *The Secret Life of Plants* (London: Penguin, 1974) p. 48.
53 Ibid. p. 19.
54 David Luke 'Parapsychology and the New Renaissance' in *A New Renaissance* ed. David Lorimer and Oliver Robinson (Edinburgh: Floris Books, 2010) p. 137.
55 Tompkins and Bird 1974, p. 19.
56 Johann Wolfgang von Goethe *Italian Journey* (New York: Schocken Books, 1968) pp. 305–06.
57 Quoted in Erich Heller *The Disinherited Mind* (New York: Farrar, Straus & Cudahy, 1957) p. 31.
58 I cannot say that I have seen the Primal Plant in the way that Goethe did, but in the Introduction to my *Rudolf Steiner* (2007 p. xxi) I write of my own experience of 'imaginative observation' in which I felt that I was not merely seeing a rose, but 'cradling' it in my consciousness.
59 From Rudolf Steiner *Goethe's Conception of the World* (London: Rudolf Steiner Press, 1928), quoted in *The Essential Steiner* ed Robert A. McDermott (San Francisco: Harper & Row, 1984) p. 49.
60 James Lovelock, *The Revenge of Gaia* (London: Penguin Books, 2007) p. 20. Lovelock points out that the use of the name Gaia was suggested to him by the novelist William Golding in 1969. Readers of Golding's novels, such as *The Lord of the Flies* (1954), in which a group of schoolboys marooned on an island

descend into savagery, may be excused for thinking that he exhibits a misanthropy similar to John Gray's, another vocal supporter of Lovelock's theories.
61 Ibid. pp. 8, 13, 17.
62 William James 'Concerning Fechner' in *The Writings of William James* ed. John J. McDermott (New York: The Modern Library, 1968) p. 540.
63 Michael Heidelberger *Nature From Within: Gustav Fechner and his Psychophysical Worldview* (Pittsburg: University of Pittsburgh Press, 2004) p. 177.
64 See Lachman 2012, pp. 82, 85, 92, 93.
65 Quoted in Walter Lowrie *Religion of a Scientist: Selections for Gustav Fechner* (New York: Pantheon Books, 1946) p. 211.
66 Ibid. p. 165.
67 Lachman 2003, p. xxvi.
68 James 1968, p. 535.
69 Lowrie 1946, p.180.
70 Ibid. p.132.
71 Ibid.p.133. On Swedenborg's 'search for the soul' see Lachman 2012 pp. 37–67.

6. The Participatory Universe

1 Steiner 2008.
2 This, in a sense, can be seen as a version of the 'turtles all the way down' theory of the universe. There are different versions of the story, but in one, the philosopher Bertrand Russell gave a lecture in which he described how the earth orbited the sun, and the sun orbited the centre of the Milky Way. At the end of the lecture an old woman stood up and told Russell that what he had just lectured on was nonsense. The earth, she said, was a flat plate lying on the back of an enormous turtle. Russell smiled at the woman and asked, 'And what is the turtle standing on?' 'You are very clever, young man' the woman said, 'very clever indeed. But I'm afraid its turtles all the way down.'
3 Ludwig Wittgenstein *Tractatus Logico-Philosophicus* (London: Routledge & Kegan Paul, 1969) p. 149.
4 Ibid.
5 A very readable and fascinating account of the uncertainties inherent in modern science can be found in Lynn Picknett and Clive Prince's *The Forbidden Universe* (London: Constable, 2011). I've already mentioned Sheldrake's *The Science Delusion*. Classics in this field are Arthur Koestler's *The Ghost in the Machine* (1967), Michael Polanyi's *Personal Knowledge* (1958), and Maslow's *The Psychology of Science* (1966). There are indeed many others but these should give readers a start.
6 Quoted in Steven Rose, ed. *From Brains to Consciousness* (London: Allen Lane, 1998) p. 247.
7 John Searle *The Mystery of Consciousness* (London: Granta Books, 1997) p. xiv.
8 Lachman 2003 pp. xxv–xxvi The original research was done by John Lorber, a specialist in hydrocephaly, in 1965, and was presented in his paper 'Hydranencephaly with Normal Development' in *Developmental Medicine and Child Neurology*, December 1965, 7: 628–633. A popular account of Lorber's work, 'Is Your Brain Really Necessary,' by Robert Lewin, appeared in the 12

December 1980 issue of *Science*.
9 The phrase 'explanatory gap' was coined in 1983 by the philosopher Joseph Levine to indicate the difficulty in explaining how physical properties produce inner experience. It is widely associated with the work of the philosopher David Chalmers.
10 Quoted in Larry Dossey 'Mind and Neurons; Consciousness and the Brain in the Twenty-First Century' in Lorimer and Robinson 2010 p. 107.
11 Quoted in Steiner 2008 p. 53.
12 Ibid.
13 Ian Marshall and Danah Zohar *Who's Afraid of Schrödinger's Cat?* (London: Bloomsbury, 1997) pp. 64–67, 249–250.
14 See my *Jung the Mystic* (New York: Tarcher/Penguin, 2010).
15 Quoted in Steiner 2008 p. 61.
16 See David Chalmers *The Conscious Mind* (Oxford: Oxford University Press, 1996).
17 Owen Barfield *History in English Words* (West Stockbridge, MA: Lindisfarne Press, 1985) p. 18.
18 Owen Barfield *Romanticism Come of Age* (Middletown, CT: Wesleyan University Press, 1986) p. 189.
19 Barfield *Saving the Appearances* pp. 116–117.
20 One suspects that this process of pairing language down has accelerated considerably since Barfield first wrote about it. Witness the flat rhetoric of much postmodern fiction, the telegraphic communiqués of text messaging (LOL, OMG) , the impatient prose of emails, and overdependence on the f-word as an all-purpose expletive.
21 Owen Barfield 'Owen Barfield and the Origin of Language' (Spring Valley, NY: St. George Publications, 1976) p. 10.
22 Lachman 2003 pp. 161–169.
23 We might think to ask how accurate a representation our consciousness provides, but this proves an exceedingly difficult question to answer, as we would somehow have to know what reality is like when we are not representing it, and here we hit a wall. The closest we come, at least according to science, are the kind of mathematical expressions I mention. If these are indeed what reality is really like, then our representation of it is an immense falsification, just as the sumptuous colours of a sunset are a 'falsification' of electromagnetic radiations, which, without the eye to interpret them, are colourless. Hence Blake's dictum that 'Where man is not, nature is barren', an insight he shared with at least one philosopher of science. Alfred North Whitehead remarked that the materialist picture of the universe revealed a nature that was 'a dull affair, soundless, scentless, colourless; merely the hurrying of material, endlessly, meaninglessly'. Agreeing with Blake, Whitehead said that in their praise of nature 'the poets are entirely mistaken. They should address their lyrics to themselves, and should turn them into odes of self-congratulation on the excellency of the human mind.' *Science and the Modern World* (New York: Macmillan, 1925) p. 77.
24 Quoted in Colin Wilson *Superconsciousness* (London: Watkins, 2009) p. 172. Paul Ricoeur *Husserl: An Analysis of His Phenomenology* (Northwestern University Press: Chicago, 1967) p.547.
25 Wheeler began using the term 'participation' in his account of the universe in 1983. Barfield used it in *Saving the Appearances* as early as 1957, although

the central theme of a continuity between the inner and outer worlds had informed his thought since 1926 and *History in English Words*. And of course, the notion that the mind and reality were in some way 'one' reaches back into our religious, philosophical, and spiritual past. The phrase 'Tat tvam asi', 'Thou art that', for example, expressing the essential unity between the human mind and the Ultimate Reality, appears in the *Chandogya Upanishad*, which is generally thought to have been written sometime in the first millennium BC. Wheeler, who, unlike Wolf Singer, had no time for the paranormal and famously called for parapsychologists to be expelled from the American Association for the Advancement of Science, is naturally considered one of the great minds of our time, and his 'participatory' notions are thought to be breakthroughs in our understanding of the universe. Barfield, who was there well ahead of him, is barely known outside of Anthroposophical circles or to readers of C.S. Lewis (he was Lewis' great friend). When I mentioned him once to an academic friend, he remarked 'You mean that Coleridge looney?' Barfield had written a book entitled *What Coleridge Thought* (1971). There is no justice in intellectual history.

26 Marshall and Zohar 1997, pp. 221–23.
27 Anthony Peake, *The Labyrinth of Time: The Illusion of Past, Present and Future* (London: Arcturus Publishing, 2012) pp. 109–11.
28 Marshall and Zohar 1997, pp. 112–13.
29 Quoted in Peake 2012, p.109.
30 Barfield 1957, 144–45.
31 I say to some extent, because although Wheeler and the scientists who follow his lead use the term 'participation' they do not share Barfield's, Wilson's and others' belief that our own volition – our will – can increase our participation. Theirs is a 'participatory universe' but, like earlier scientific models, it is a passive one. That our consciousness participates in the universe is for them as much a 'law' as is gravity. For people like Barfield, the idea is to *consciously increase* our participation, and hence freedom. And by consciously increasing our participation, we do not then gain greater control over the things that make up our world, but of our *picture* of that world, just as while I cannot control what takes place in a television programme I am watching, I can control the quality of the picture on the screen; i.e., increase or decrease the focus, clarity, and so on.
32 Barfield 1957 p. 160.
33 See Lachman 2010, p.241 n31, p. 242 n41, pp. 246–47 n13.
34 Colin Wilson *Beyond the Occult* (London: Bantam Press, 1988) p. 327.
35 Coleridge's friend, Thomas De Quincey, the 'English Opium Eater' added to this insight when he once remarked, apropos of opium's power to elicit dreams, that if a man's head was full of oxen, he would dream of oxen too. Meaning that the drug itself – whether opium, LSD, or anything else – did not provide poetic or visionary dreams, but only released what was already in the drug taker's mind. De Quincey's drug experiences are interesting to read about because his head was full of literature, philosophy, art, history and not oxen, something that is true of Aldous Huxley and other brilliant writers who have also written about their drug experiences. Their drug experiences produced literature, because they had the material for literature already in their heads. In heads not so well furnished it may produce merely an inarticulate 'wow'.
36 Faust Part One. *Die Geisterwelt ist nicht verschlossen; Dein Sinn ist zu, dein Herz ist tot!*

37 Ernst Cassirer *An Essay on Man* (New Haven: Yale University Press, 1944) p.24.
38 Ibid. p. 3.
39 Ibid. p. 5.
40 Ibid. p. 9.
41 Ibid. p. 195.
42 Ibid. p. 114.
43 Ibid. p. 111.
44 Ibid. p. 143.

7. As Far as Thought Can Reach

1 Cassirer 1944, p. 22.
2 Scheler 1928, pp. 5–6, quoted in Cassirer Ibid.
3 Quoted in William Barrett *What is Existentialism?* (New York: Grove Press, 1965) p. 114.
4 Cassirer 1944 p. 9.
5 Ibid. p. 9.
6 Berdyaev 1937 p. 177.
7 Ibid. p. 179.
8 Friedrich Nietzsche *Thus Spoke Zarathustra* in *The Portable Nietzsche* translated and edited by Walter Kaufmann (New York: Penguin, 1977) p. 130.
9 In *The Order of Things* (New York: Pantheon, 1970) and *The End of History and the Last Man* (New York: Free Press, 1992) respectively.
10 John Horgan *The End of Science* (London: Abacus, 1998).
11 Steiner 2001 pp. 1–2.
12 See my article '2013; Or What to do When the Apocalypse Doesn't Arrive' http://www.disinfo.com/2009/10/2013-or-what-to-do-when-the-apocalypse-doesn%E2%80%99t-arrive/
13 Bertrand Russell *Why I am Not a Christian and Other Essays on Religion and Related Subjects* (New York: Simon & Schuster, 1986) pp. 32–33.
14 In 'The Hollow Men' Eliot wrote : *This is the way the world ends/Not with a bang but a whimper.*
15 Gary Lachman 'Giant Jupiters May Be Rare, Says UCLA Astronomer. Our Solar System May Be Special After All' *The Basics: Newsletter of the Life & Physical Sciences From the UCLA College of Letters and Science* Vol 1, No. 2 Spring 1995.
16 Marshall and Zohar 1997 pp. 43–44.
17 Svante Arrhenius (1859–1927) was a Swedish chemist and physicist. He received the Nobel Prize for Chemistry in 1903. For Arrheniuis' appreciation of Swedenborg, and for details about Swedenborg's scientific work, see Lachman 2012 pp. 48–49.
18 Marshall and Zohar 1997 p. 44.
19 Rainer Maria Rilke *Duino Elegies* trans. Stephen Mitchell in Mitchell 1984 p. 199. Rilke's 'just once, no more' has clear resonances with Nietzsche's notion of the eternal recurrence, the idea that *this* life, and our actions and choices within it, is fated to be repeated again and again without change, for eternity. If things do recur exactly as they happen now, and if they recur eternally,

then *this* recurrence is exactly the same as the last and the next. And as there is no appreciable difference between recurrences, in effect all we have is *now*, Rilke's 'just once'. Our response to the eternal recurrence of our lives was for Nietzsche the test of whether or not we had the potential to become 'overman', whether or not we can embrace our fate, what Nietzsche called *amor fati* or 'love of fate'. For an excellent essay on the links between Rilke and Nietzsche see 'Rilke and Nietzsche with a Discourse on Thought, Belief, and Poetry' in Erich Heller, *The Importance of Nietzsche* (Chicago: University of Chicago Press, 1988) pp. 87–126.
20 Rainer Maria Rilke *Duino Elegies* trans. J.B. Leishman and Stephen Spender (New York; W.W. Norton & Co., 1939) p. 129.
22 Ibid.
23 Ibid. pp. 128–29.
24 Sir Arthur Eddington *The Nature of the Physical World* (Whitefish, MO: Kessinger Publications, 2005) p. 74.
25 'Schrödinger's cat' refers to a thought experiment, found in many books on quantum physics, in which a cat is locked into a box with a flask of hydrocyanic acid and a Geiger counter equipped with a small piece of radioactive material. If the radioactive material decays, the Geiger counter detects it and sends a message to a mechanical device that will smash the flask, releasing the poison and killing the cat. If it doesn't decay, nothing will happen. Quantum theory says that at a certain point, quantum indeterminacy tells us that the cat is both alive and dead simultaneously, or rather an indefinite mix of the two. It is not until the experimenter opens the box and looks in that the cat is definitely one or the other. The observer's 'participation' with what he is observing 'collapses the wave function' (as in the twin slit experiment) and 'decides' which it will be. For an account of the experiment see F. David Peat *The Philosopher's Stone* (New York: Bantam, 1991) pp. 68–72.
26 See Ilya Prigogine and Isabelle Stengers' *Order Out of Chaos* (London: Flamingo, 1997).
27 http://129.81.170.14/~tipler/summary.html
28 On this note it might be helpful to ask why we never wonder why there is good in the world. We always ask 'Why is there evil?', but we never ask 'Why is there good?' This is because we assume good to be the 'default setting', as it were, the bottom line foundation, and evil something that developed out of this, as a corruption of it. But it strikes me as much easier to see a world in which evil – pain, suffering, death – were fundamental, and good something new. The natural world is one in which many of the goods we cherish do not obtain. We know that at that level, it is a 'dog eat dog' world in which only the strong survive. In that world might does make right. It doesn't take much imagination to transfer this world to the human one – indeed, many would say we don't need to transfer it at all. But if this is the case, then good would be something very much out of place and in need of protection. I would say there is good in the world because it is little bits of God or whatever you want to call some idea of absolute value seeping through the cracks of a world, not necessarily evil, but one initially void of good. Good, then, has a history, and I would say it does not appear until human beings began to actualise it.
29 Scheler 1978 p. 70.
30 Scheler 1979 p. 32.

Selected Bibliography

Barfield, Owen (1957) *Saving the Appearances,* New York: Harcourt Brace & World.
—, (1976) 'Owen Barfield and the Origin of Language,' Spring Valley, NY: St. George Publications.
—, (1985) *History in English Words,* West Stockbridge, MA: Lindisfarne Press.
—, (1986) *Romanticism Come of Age,* Middletown, CT: Wesleyan University Press.
Barrett, William (1965) *What is Existentialism?* New York: Grove Press.
Barzun, Jacques (1974) *The Use and Abuse of Art,* Princeton: Princeton University Press.
Benjamin, Walter (1978) *Reflections,* New York: Harcourt, Brace, Jovanovich.
—, (1992) *Illuminations,* London: HarperCollins.
Benz, Ernst (1983) *The Mystical Sources of German Romantic Philosophy,* Allison Park, PA: Pickwick Publications.
Berdyaev, Nikolai (1937) *The Destiny of Man,* London: Geoffrey Bly.
—, (1962) *The Meaning of the Creative Act,* New York: Collier Books.
Bergson, Henri (1912) *Introduction to Metaphysics,* New York: G.P. Putnam's Sons.
Blake, William (1982) *The Complete Poetry and Prose of William Blake,* ed. David V. Erdman, Berkeley: University of California Press.
Burckhardt, Jacob (1944) *The Civilization of the Renaissance in Italy,* London: Phaidon Press.
Butler, Samuel (1917) *The Notebooks of Samuel Butler,* New York: E.P. Dutton & Co.
Cassirer, Ernst (1944) *An Essay on Man,* New Haven: Yale University Press.
—, (1972) *The Individual and the Cosmos in Renaissance Philosophy,* Philadelphia: University of Pennsylvania Press.
Caws, Peter (1999) *Sartre,* London: Routledge.
Chalmers, David (1996) *The Conscious Mind,* Oxford: Oxford University Press.
Crick, Francis (1994) *The Astonishing Hypothesis: The Scientific Search for the Soul,* London: Simon & Schuster.
Damon, S. Foster (1979) *A Blake Dictionary,* Boulder, CO: Shambhala.
de Beauvoir, Simone (1992) *The Prime of Life,* New York: Perseus Books.

de Lubicz, R.A. Schwaller (1982) *Nature Word,* West Stockbridge: Lindisfarne Press.
—, (1985) *Esotericism and Symbol,* New York: Inner Traditions.
Derleth, August (1969) ed. *Tales of the Cthulhu Mythos,* Sauk City, WI: Arkham House.
Drob, Sanford L. (2000) *Kabbalistic Metaphors,* Northvale, NJ: Jason Aronson Inc.
Eddington, Sir Arthur (2005) *The Nature of the Physical World,* Whitefish, MO: Kessinger Publications.
Foucault, Michel (1970) *The Order of Things,* New York: Pantheon.
Forbes, Elliot (1991) *Thayer's Life of Beethoven, Vol. I,* Princeton: Princeton University Press.
Frankl, Victor (1984) *Man's Search for Meaning,* New York: Simon & Schuster.
Fukuyama, Francis (1992) *The End of History and the Last Man,* New York: Free Press.
Goethe, Johann Wolfgang von (1968) *Italian Journey,* New York: Schocken Books.
Goodrick-Clarke, Nicholas (1998) *Hitler's Priestess: Savitri Devi, the Hindu Aryan Myth and Neo-Nazism,* New York: NYU Press.
Gray, John (2002) *Straw Dogs: Thoughts on Humans and Other Animals,* London: Granta.
—, (2011) *The Immortalization Commission,* London: Penguin.
Heidelberger, Michael (2004) *Nature from Within: Gustav Fechner and his Psychophysical Worldview,* Pittsburgh: University of Pittsburgh Press.
Heller, Erich (1957) *The Disinherited Mind,* New York: Farrar, Straus, & Cudahy.
—, (1988) *The Importance of Nietzsche,* Chicago: University of Chicago Press.
Hesse, Hermann (1995) *The Journey to the East,* London: Picador.
Hillman, James (1977) *Re-Visioning Psychology,* New York: Harper Colophon Books.
Hoffman, Edward (1988) *The Right To Be Human: A Biography of Abraham Maslow,* New York: Tarcher.
Hoffmann, E.T.A. (1982) *Tales of Hoffmann,* R.J. Hollingdale trans. London: Penguin.
Horgan, John (1998) *The End of Science,* London: Abacus.
Huxley, Julian (1957) *New Bottles for New Wine,* London: Chatto & Windus, 1957.
James, William (1968) *The Writings of William James,* ed. John J. McDermott, New York: The Modern Library.
Joshi, S.T. (1990) *The Weird Tale,* Austin, TX: University of Texas Press.
—, (1996) *H.P. Lovecraft: A Life,* West Warwick, RI: Necronomicon Press.
Jung, C.G. (1991) *Collected Works,* Volume 17, London: Routledge.
Kazantzakis, Nikos (1960) *The Saviours of God: Spiritual Exercises,* New York: Simon & Schuster.
Koestler, Arthur (1967) *The Ghost in the Machine,* New York: Macmillan.

Lachman, Gary (2003) *Turn Off Your Mind: The Mystic Sixties and the Dark Side of the Age of Aquarius,* New York: Disinformation.
—, (2003) *A Secret History of Consciousness,* Great Barrington, MA: Lindisfarne.
—, (2003) *The Dedalus Book of the Occult: A Dark Muse,* Sawtry, Cambs: Dedalus.
—, (2004) *The Dedalus Occult Reader: The Garden of Hermetic Dreams,* Sawtry, Cambs: Dedalus.
—, (2006) *In Search of P.D. Ouspensky: The Genius in the Shadow of Gurdjieff,* Wheaton, Ill: Quest Books.
—, (2007) *Rudolf Steiner: An Introduction to His Life and Work,* New York: Tarcher/Penguin; Edinburgh: Floris Books.
—, (2008) *The Dedalus Book of Literary Suicides: Dead Letters,* Sawtry, Cambs: Dedalus Ltd.
—, (2008) *Politics and the Occult,* Wheaton, Ill: Quest Books.
—, (2010) *The Dedalus Book of the 1960s: Turn Off Your Mind,* Sawtry, Cambs: Dedalus.
—, (2010) *Jung the Mystic* New York: Tarcher/Penguin.
—, (2011) *The Quest for Hermes Trismegistus,* Edinburgh: Floris Books.
—, (2012) *Swedenborg: An Introduction to His Life and Ideas,* New York: Tarcher/Penguin.
Lindsay, David (1987) *The Haunted Woman,* Edinburgh: Canongate.
—, (1992) *A Voyage to Arcturus,* Edinburgh: Canongate.
Lorimer, David & Robinson, Oliver (2010) eds. *A New Renaissance,* Edinburgh: Floris Books.
Lovecraft, H.P. (1968) *Selected Letters 1925–1929,* Sauk City, WI: Arkham House.
Lovelock, James (2007) *The Revenge of Gaia,* London: Penguin Books.
Lowrie, Walter (1946) *Religion of a Scientist: Selections for Gustav Fechner,* New York: Pantheon Books.
Marshall, Ian & Zohar, Danah (1997) *Who's Afraid of Schrödinger's Cat?* London: Bloomsbury.
Maslow, Abraham (1976) *The Father Reaches of Human Nature,* New York: Penguin.
—, (1996) *Future Visions: The Unpublished Papers of Abraham Maslow,* ed. Edward Hoffman, Thousand Oaks, CA: Sage Publications.
—, (1998) *Maslow on Management,* Chichester, NY: John Wiley.
Matt, Daniel C. (1996) *The Essential Kabbalah,* New York: Harper Collins.
McDermott, Robert A. (1984) ed. *The Essential Steiner,* San Francisco: Harper & Row.
McGilchrist, Iain (2010) *The Master and His Emissary,* New Haven: Yale University Press.
Milosz, O.V. de Lubicz (1985) *The Noble Traveller: The Life and Writings of O.V. de L. Milosz,* ed. Christopher Bamford, West Stockbridge, MA: Lindisfarne.
Monod, Jacques (1971) *Chance and Necessity,* New York: Vintage Books, 1971.

Moskvitin, Jurij (1974) *An Essay on the Origin of Thought,* Columbus: Ohio University Press.
Nietzsche, Friedrich (1967) *The Will to Power,* ed. Walter Kaufmann, New York: Random House.
—, (1977) *The Portable Nietzsche,* translated and edited by Walter Kaufmann, New York: Penguin.
Ouspensky, P.D. (1970) *The Fourth Way,* New York: Alfred A. Knopf.
Pagels, Elaine (1979) *The Gnostic Gospels,* New York: Random House.
Parry, Idris, & Keegan, Paul (nd) eds. *Essays on Dolls,* London: Penguin.
Peake, Anthony (2012) *The Labyrinth of Time: The Illusion of Past, Present and Future* London: Arcturus Publishing.
Peat, F. David (1991) *The Philosopher's Stone,* New York: Bantam.
Penrose, Roger (2005) *The Road to Reality,* New York: Knopf.
Picknett, Lynn & Prince, Clive (2011) *The Forbidden Universe,* London: Constable.
Pinker, Steven (1997) *How the Mind Works,* New York: Norton.
Prigogine, Ilya & Stengers, Isabelle (1997) *Order Out of Chaos,* London: Flamingo.
Raine, Kathleen (1982) *The Inner Journey of the Poet,* New York: George Braziller.
—, (1985) *Defending Ancient Springs,* West Stockbridge, MA: Lindisfarne.
Rilke, Rainer Maria (1939) *Duino Elegies,* trans. J.B. Leishman and Stephen Spender, New York; W.W. Norton & Co.
—, (1984) *The Selected Poetry of Rainer Maria Rilke,* ed. and trans. Stephen Mitchell, New York: Vintage Books.
—, (1985) *The Notebooks of Malte Laurids Brigge,* trans. Stephen Mitchell, New York: Vintage Books.
Robinson, James (2000) ed. *The Nag Hammadi Library in English,* New York: HarperOne.
Rose, Steven (1998) ed. *From Brains to Consciousness,* London: Allen Lane.
Russell, Bertrand (1986) *Why I am Not a Christian and Other Essays on Religion and Related Subjects,* New York: Simon & Schuster.
Salaman, Clement (2007) trans. *Asclepius, The Perfect Discourse of Hermes Trismegistus* London: Duckworth.
Sartre, Jean Paul (2000) *Being and Nothingness,* Abingdon: Routledge.
—, (1975) *Nausea,* Harmondsworth: Penguin Books.
Scheler, Max (1978) *Man's Place in Nature,* New York: Noonday Press.
—, (1979) *The Nature of Sympathy,* London: Routledge & Kegan Paul.
—, (1980) *Problems of a Sociology of Knowledge,* London: Routledge & Kegan Paul.
Scholem, Gershom (1965) *On the Kabbalah and its Symbolism,* London: Routledge & Kegan Paul.
Schreck, Nicholas (1988) ed. *The Manson File,* New York: Amok Press.
Searle, John (1997) *The Mystery of Consciousness,* London: Granta Books.
Sennett, Richard (2008) *The Craftsman,* London: Allen Lane.

Sheldrake, Rupert (2012) *The Science Delusion,* London: Coronet.
Sherrard, Philip (1992) *Human Image: World Image,* Ipswich: Golgonooza Press.
Smoley, Richard (2006) *Forbidden Faith: The Gnostic Legacy,* New York: Harper Collins.
Solovyov, Vladimir (1985) *The Meaning of Love,* West Stockbridge, MA: Inner Traditions/Lindisfarne.
Spiegelberg, Herbert (1976) *The Phenomenological Movement,* Vol. 1, The Hague: Martinus Nijhoff.
Staude, John Raphael (1967) *Max Scheler,* New York: The Free Press.
Steiner, George (1978) *Has Truth a Future?* London: BBC.
—, (1987) *Martin Heidegger,* Chicago: University of Chicago Press.
—, (2001) *Grammars of Creation,* London: Faber and Faber.
—, (2008) ed. *Is Science Nearing its Limits?* Manchester: Carcanet.
Swedenborg, Emanuel (1988) *The True Christian Religion,* London: Swedenborg Society.
Tompkins, Peter, & Bird, Christopher (1974) *The Secret Life of Plants,* London: Penguin, 1974.
Van Dusen, Wilson (1975) *The Presence of Other Worlds,* London: Wildwood House.
Viereck, Peter (2003) *Metapolitics: The Roots of the Nazi Mind,* Piscataway, NJ: Transaction Publishers.
Waite, A. E. (1901) *The Life of Louis Claude de Saint-Martin,* London: Philip Wellby.
Wells, H.G. (1934) *Experiment in Autobiography,* London: Gollancz, 1934.
—, (2005) *The War of the Worlds,* London: Penguin.
—, (2005) *A Modern Utopia,* London: Penguin Books.
Wilson, Colin (1959) *The Stature of Man,* Boston: Houghton Mifflin.
—, (1962) *The Strength to Dream,* Cambridge: Houghton Mifflin.
—, (1965) *Beyond the Outsider,* Boston: Houghton Mifflin.
—, (1970) *Poetry and Mysticism,* San Francisco: City Lights.
—, (1979) *The Haunted Man: The Strange Genius of David Lindsay,* San Bernadino, CA: Borgo Press.
—, (1988) *Beyond the Occult,* London: Bantam Press.
—, (1998) *Alien Dawn,* London: Virgin Publishing.
—, (1998) *Below the Iceberg,* San Bernardino: The Borgo Press.
—, (2009) *Superconsciousness,* London: Watkins.
Whitehead, Alfred North (1925) *Science and the Modern World,* New York: Macmillan.
—, (1928) *Symbolism: Its Meaning and Effect,* Cambridge: Cambridge University Press.
—, (1948) *Essays in Science and Philosophy,* London: Rider.
Wittgenstein, Ludwig (1961) *Tractatus Logico-Philosophicus,* London: Routledge & Kegan Paul.
Wollstonecraft, Mary (1987) *Letters Written During a Short Residence in Sweden, Norway, and Denmark,* Harmondsworth: Penguin.

Index

Adams, Douglas 172
Andersen, Hans Christian 68
Appleyard, Bryan 48
Arnim, Bettina von 68
Arnold, Matthew 212
Arrhenius, Svante 215
Asclepius 16

Baal Shem Tov 76
Bach, Johann Sebastian 57, 119
Backster, Cleve 164, 169, 180, 186, 224
Bacon, Francis 154
Ballard, J.G. 48
Barfield 149, 151f, 183–86, 188, 190, 192–95, 201f
Barfield, Owen 144, 148, 162, 182
Barzun, Jacques 58, 211
Beauvoir, Simone de 42
Becker, Ernst 52
Beethoven, Ludwig van 46, 68, 71, 86f
Benedict, Ruth 130
Benjamin, Walter 34
Berdyaev, Nikolai 17, 58f 108–11, 113, 138f, 163, 176, 202f, 206, 209f, 221
Bergson, Henri 84–86, 88, 95, 99, 114, 137, 147, 170, 180, 182, 201, 220f
Bird, Christopher 164f, 168
Blake, William 32, 62, 64f, 72, 80–83, 142, 150, 155–59, 161, 163f, 166, 189, 195, 199–201, 210, 221
Boehme, Jacob 30
Bohm, David 30
Bohr, Niels 84
Borges, Jorge Luis 85
Breton, André 58
Bruno, Giordano 61

Bryusov, Valery 132f
Buber, Martin 128
Bucke, R.M. 72
Buddha 46
Burckhardt, Jacob 143
Burke, Edmund 145f
Burrough, Edgar Rice 72
Butler, Samuel 100f, 104

Cabanis, Pierre 176
Cage, John 57
Carter, John 72
Cassirer, Ernst 17, 143, 201–03, 206–09, 221
Chalmers, David 170, 180
Cioran, Emil 51
Clausius, Rudolf 211f
Coleridge, Samuel Taylor 144f, 199
Copernicus 60f, 149
Corbin, Henry 65
Cordovero, Moses 31
Crick, Francis 54, 81, 171, 175, 177, 189, 216

Dante 63
Darwin 206
Dawkins, Richard 48, 53
Democritus 43
Dennett, Daniel 176f
Descartes, René 61, 154
Dick, Philip K. 14
Dostoyevsky 133
Drob, Sanford L. 35, 90, 92
Dusen, Wilson Van 101
Dylan, Bob 14, 162

Eckhart, Meister 29

Eddington, Sir Arthur 188, 220
Einstein, Albert 46, 64f, 168, 180
Elijah 30f
Eliot, T.S. 185, 212
Ellenberger, Henri 168
Emerson, Ralph Waldo 183

Falk, Rabbi Samuel Jacob Hayyim 108
Fechner, Gustav 26, 167, 169f, 222
Fichte, Johann 134
Foreman, Dave 50
Foucault, Michel 211
Fox, Gardner 67
Frankl, Victor 96–99, 103f, 135
Franz, Marie-Louise Von 27
Freud, Sigmund 53, 127, 168, 206
Fromme, Lynette 'Squeaky' 49
Fukuyama, Francis 211

Galileo Galilei 87f
Gasset, Ortega y 111
Gebser, Jean 147–49, 151–54, 156, 160, 182, 193, 195, 201
Gebser, Jean 147, 193
Gell-Mann, Murray 39
Gödel, Kurt 174f
Goethe, Johann Wolfgang von 26, 63, 68f, 110, 145, 165f, 170, 200
Goodrick-Clarke, Nicholas 50
Good, Sandra 49
Gott, J.Richard III 37
Gray, John 23, 48–55, 106, 154, 203, 216
Gribbin, John 214
Gurdjieff, G.I. 14, 93, 169
Guth, Alan 36

Hardenberg, Georg Friedrich Phillipp von 74
Hawking, Stephen 38, 48, 56, 155, 172, 175, 203
Hegel 30, 63, 80, 83, 116
Heidegger, Martin 14, 30, 76f, 83, 111, 153, 160–162, 203, 207, 209, 218
Heisenberg, Werner 189
Heller, Erich 82, 183
Hermes Trismegistus 11, 16, 23, 25, 30
Hesse, Hermann 14, 75, 100, 104

Hillman, James 143
Hitler, Adolf 208
Hoffmann, E.T.A 67–71, 74
Houellebecq, Michel 106f, 109
Howard, Robert E. 43
Hoyle, Fred 46
Hulewicz, Witold von 218
Humphrey, Nicholas 176f
Husserl, Edmund 73, 77, 111, 114, 150, 188, 207
Huxley, Aldous 51f, 132, 134, 146
Huxley, Julian 51, 95, 130, 195, 202

James, William 72, 133f, 167, 170, 179
Jarry, Alfred 58
Johnson, Dr Samuel 144–46
Joshi, S.T. 48
Joyce, James 39
Jung, C.G. 14, 27, 120f, 128f, 180, 196f
Juric, Mario 37

Kadmon, Adam 32, 65, 75, 107
Kafka, Franz 43
Kauffman, Stuart 178
Kazantzakis, Nikos 20, 93–96, 101, 104, 138, 210, 222
Kierkegaard, Søren 15
Kleist, Heinrich von 162f, 193
Kundera, Milan 217

LaBianca, Leno 49
—, Rosemary 49
Lambert, Angela 106
Laplace, Pierre-Simon 155
Leibniz, Gottfried Wilhelm 90
Lem, Stanislaw 169
León, Moses de 30
Lessing, Doris 18
Leucippus 43
Lichtenberg, Georg Christoph 168
Lindsay, David 69—71, 96, 222
Locke, John 121f
Longinus 145
Lovecraft, H.P. 23, 42–47, 50, 52f, 61, 98, 106f, 117, 137, 215
Lovelock, James 48, 167, 176
Lubicz, René Schwaller de 83f, 86, 88, 99, 103, 182, 186

Luke, David 165
Luria, Isaac 19f, 30–32, 34

Malik, Kenan 50
Manson, Charles 49f
Marx, Karl 206
Maslow, Abraham 21f, 24f, 51, 77, 93, 102, 104, 116, 120–31, 134–38, 163, 195f, 222
Matisse, Henri 84
Matt, Daniel C. 100
McGilchrist, Iain 58, 77–82, 83, 85, 88f, 99, 112, 141f, 180
Merleau-Ponty, Maurice 111
Mersenne, Marin 154
Milosz, O.V. de Lubicz 62–66, 75, 157
Mirandola, Pico della 150, 207
Monod, Jacques 41–43, 52, 98, 117, 203, 214
Mother Teresa 46
Müller, Max 184–87

Neumann, Erich 59
Newton, Isaac 36, 63, 65
Nietzsche, Friedrich 50, 61, 80, 94, 102, 116, 134, 162, 183, 206, 210, 211
Novalis 69, 75

Offenbach, Jaques 68
Orwell, George 132
Ouspensky, P.D. 72, 199

Parry, Sir Hubert 156
Pascal, Blaise 39, 61–64, 151, 153f, 159
Peake, Anthony 190
Penrose, Roger 57
Petrarch (Francesco Petrarca) 143f, 149, 151, 165
Pinker, Steven 68
Plato 31, 43, 46, 63, 76, 107, 136, 154, 156
Poe, Edgar Allan 43, 68
Polanyi, Michael 82
Pope, Alexander 144
Pope John Paul II 111
Popper, Karl 173f
Proyas, Alex 14
Pynchon, Thomas 14

Pythagoras 57, 59f, 68, 203

Rachel Carson 140
Raine, Kathleen 50, 55, 58, 72, 82, 156, 158
REM 211
Rembrandt, Harmenszoon van Rijn 119
Ricoeur, Paul 188
Rilke, Rainer Maria 146f, 161, 218f
Rousseau, Jean Jacques 145
Russell, Bertrand 212, 220

Sade, Marquis de 49
Saint-Martin, Louis Claude de 20f, 22f, 25, 75f, 103f, 109, 157–61, 163, 196, 201, 203, 210, 218
Salome, Lou-Andreas 94
Sartre, Jean Paul 42–44, 46, 49, 52, 61, 80, 86, 104, 117, 121–23, 193, 198
Scheler, Max 17, 26, 93, 104, 111–20, 122, 126, 128, 135, 137f, 162, 166, 186, 193, 201–03, 206–08, 221–24
Schelling, Friedrich von 169
Schoenberg, Arnold 57
Scholem, Gershom 33, 100, 204
Schopenhauer, Arthur 49
Schrödinger, Erwin 220
Schwartz, Julius 67
Searle, John 176f, 179f, 189, 203
Sekowsky, Mike 67
Sennett, Richard 103
Sex Pistols 211
Shaw, Bernard 200
Sheldrake, Rupert 172, 176
Sherrington, Charles 179
Sinclair, Upton 178
Singer, Wolf 179–81
Smith, Clark Ashton 43
Socrates 31
Solovyov, Vladimir 109f, 113, 167
Spiegelberg, Herbert 114
Stalin, Joseph 103, 208
Stapledon, Olaf 46
Stark, Werner 117
Staude, John Raphael 113
Stein, Edith 111
Steiner, George 38, 77, 120, 125, 161, 174, 211

Steiner, Rudolf 19, 152, 166, 183, 185, 193–95, 198, 201, 219
Steinsaltz, Rabbi Adin 90f, 97, 103, 104, 134
Stekel, Wilhelm 93–95
Strange, Adam 66f, 69
Stravinsky, Igor 57
Swedenborg, Emanuel 20, 25, 32, 62, 64f, 82, 101, 103–05, 107f, 156–58, 168, 170, 201, 215, 221

Tallis, Raymond 50
Tarkovsky, Andrei 169
Tate, Sharon 49
Tchaikovsky, Pyotr Ilyich 68
Tipler, Frank 221
Tompkins, Peter 164f, 168
Toynbee, Arnold 137

Uexküll, Jakob Johan von 73

Van Dusen 103f
Vinci, Leonardo da 46
Virgil 142
Vital, Chayyim 31
Voltaire 90

Waite, A.E. 158
Wallace, Alfred 52
Watson, John 122
Weinberg, Steven 39, 59, 66, 154, 172
Weir, Peter 14
Wells, H.G. 46, 125f, 130–32, 134, 136, 138f, 142, 146, 202
Wheeler, John Archibald 190–92, 204
Whitehead, Alfred North 86–88, 99, 157, 170, 180, 199, 201, 207, 221
Whitman, Walt 27
Wilde, Oscar 62
Wilson, Colin 15, 20f, 24, 55, 69, 74, 83, 107, 137, 147, 150, 195, 197
Wittgenstein 83, 175
Wollstonecraft, Mary 145f
Wordsworth, William 117, 144–46, 157

Yeats, W.B. 78
Yohai, Rabbi Shimon Bar 30

Zevi, Sabbatai 108
Zinzendorf, Count 108
Zohar, Danah 214, 216

Other books by this author

The Quest for Hermes Trismegistus
From Ancient Egypt to the Modern World

Gary Lachman

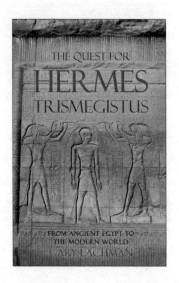

'Immensely erudite'
— David Lorimer, Scientific and Medical Network Review

From the sands of Alexandria and the Renaissance palaces of the Medicis, to our own postmodern times, this spiritual adventure story traces the profound influence of Hermes Trismegistus – the 'thrice-great one', as he was often called – on the western mind.

In this book, the author brings to life the mysterious character of this great spiritual guide, exposing the many theories and stories surrounding him, and revitalizing his teachings for the modern world.

Also available as an eBook.

florisbooks.co.uk

Rudolf Steiner
An Introduction to His Life and Work

Gary Lachman

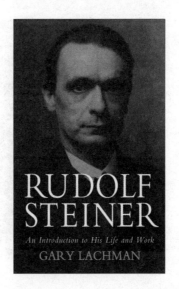

Rudolf Steiner – educator, architect, artist, philosopher and agriculturalist – ranks amongst the most creative and prolific figures of the early twentieth century. Yet he remains a mystery to most people. This is the first truly popular biography of the man behind the ideas, written by a sympathetic but critical outsider.

The author tells Steiner's story lucidly and with great insight. He presents Steiner's key ideas in a readable, accessible way, tracing his beginning as a young intellectual in the ferment of *fin de siècle* culture to the founding of his own metaphysical teaching, called anthroposophy.

This book is a full-bodied portrait of one of the most original philosophical and spiritual luminaries of the last two centuries.

florisbooks.co.uk

A Secret History of Consciousness

Gary Lachman

In this book, the author argues that consciousness is not a result of neurons and molecules, but is actually responsible for them. Meaning, he proposes, is not imported from the outer world, but rather creates the world. He shows that consciousness is a living, evolving presence whose development can be traced through different historical periods.

Concentrating on the late nineteenth-century onwards, Lachman exposes the 'secret history' of consciousness through thinkers such as P. D. Ouspensky, Rudolf Steiner, and Colin Wilson, as well as more mainstream philosophers like Henri Bergson, William James, Owen Barfield and psychologist Andreas Mavromatis.

This is a far-reaching book from an exciting contemporary thinker.

florisbooks.co.uk